EAI/Springer Innovations in Communication and Computing

Series editor
Imrich Chlamtac, European Alliance for Innovation, Ghent, Belgium

Editor's Note

The impact of information technologies is creating a new world yet not fully understood. The extent and speed of economic, life style and social changes already perceived in everyday life is hard to estimate without understanding the technological driving forces behind it. This series presents contributed volumes featuring the latest research and development in the various information engineering technologies that play a key role in this process.

The range of topics, focusing primarily on communications and computing engineering include, but are not limited to, wireless networks; mobile communication; design and learning; gaming; interaction; e-health and pervasive healthcare; energy management; smart grids; internet of things; cognitive radio networks; computation; cloud computing; ubiquitous connectivity, and in mode general smart living, smart cities, Internet of Things and more. The series publishes a combination of expanded papers selected from hosted and sponsored European Alliance for Innovation (EAI) conferences that present cutting edge, global research as well as provide new perspectives on traditional related engineering fields. This content, complemented with open calls for contribution of book titles and individual chapters, together maintain Springer's and EAI's high standards of academic excellence. The audience for the books consists of researchers, industry professionals, advanced level students as well as practitioners in related fields of activity include information and communication specialists, security experts, economists, urban planners, doctors, and in general representatives in all those walks of life affected ad contributing to the information revolution.

Indexing: This series is indexed in Scopus, Ei Compendex, and zbMATH.

About EAI

EAI is a grassroots member organization initiated through cooperation between businesses, public, private and government organizations to address the global challenges of Europe's future competitiveness and link the European Research community with its counterparts around the globe. EAI reaches out to hundreds of thousands of individual subscribers on all continents and collaborates with an institutional member base including Fortune 500 companies, government organizations, and educational institutions, provide a free research and innovation platform.

Through its open free membership model EAI promotes a new research and innovation culture based on collaboration, connectivity and recognition of excellence by community.

More information about this series at http://www.springer.com/series/15427

Honghao Gao • Yuyu Yin

Editors

Intelligent Mobile Service Computing

 Springer

Editors
Honghao Gao
Shanghai University
Shanghai, China

Yuyu Yin
Hangzhou Dianzi University
Hangzhou, China

ISSN 2522-8595   ISSN 2522-8609 (electronic)
EAI/Springer Innovations in Communication and Computing
ISBN 978-3-030-50186-0   ISBN 978-3-030-50184-6 (eBook)
https://doi.org/10.1007/978-3-030-50184-6

This Springer imprint is published by the registered company Springer Nature Switzerland AG.
The registered company address is: Gewerbestrasse 11, 6330 Cham, Switzerland

Preface

The Intelligent Service Computing in Mobile Environment

With the rapid development of 5G technologies, mobile terminal devices with edge computing capacity are prevailing, and the ubiquitous availability of mobile services that enable software applications can be guaranteed on quality in the wireless network environment. The intelligence-empowered mobile computing has significantly changed the style of our life, study, and work. Along with this, a large number of new-style intelligent services have emerged, with features of complexity, diversity, and crossover. At this point, the intelligent service computing in the mobile environment is becoming the focus of the next-generation computer network, which will be an important Internet infrastructure for interpersonal connection that links people and the physical world.

How to utilize the research of Artificial Intelligence and Big Data to prompt the research of new theories, methods, and applications of service computing has become a hot research point. In recent years, deep learning has made significant breakthroughs in many areas of Artificial Intelligence, including CNN, RNN, and GAN. However, these traditional models need to upload and handle data at the cloud end, which cannot meet the requirements of timeliness and mobility. We aim to point out that mobile service computing shall pay attention to the smart processing by using 5G, mobile services, and the Internet of Things (5MIoT). We have a beautiful vision that the breakthrough technology in these fields can enable mobile users to enjoy high-quality service experience with high performance, low latency, and high bandwidth.

This book includes six chapters focusing on many issues including security model, performance and energy efficiency, formal verification/testing/inspection, QoS edge computing, electricity data of smart meters, and electricity data of smart meters in mobile environment.

The first chapter, titled "A Cross-Domain Security Model Based on Internet of Vehicles," points out that with the continuous expansion of application scopes of the Internet of Vehicles, attacks have increased correspondingly. There have been too many attack accidents in IoV. Now, the intelligent vehicles have become important targets of the hacker. The authors look forward to constructing a cross-domain security model of IoV. First, they introduce the security domain and analyze its divisions. Second, they study the structure of the cross-domain of IoV. Finally, they divide the security domain into three layers, namely cross-domain application layer, cross-domain network layer, and cross-domain perception layer. Also, they study the trust relationship in IoV and make use of the method of trust evaluation to ensure transaction security and construct a trustworthy environment for cross-domain. Their scheme can partly ensure the security of communications and the privacy of vehicles.

The second chapter, titled "A Framework to Improve Performance and Energy Efficiency of Embedded Intelligence Service Systems," discusses that due to the high real-time requirement and limited resources, the computing performance and power consumption are critical to embedded intelligent service systems. The thread group scheduling strategy, as a common method to optimize the performance of the multi-core processor, performs poorly in power optimization. However, as data volume continues proliferating, the power consumption keeps increasing. Correspondingly, many power optimization methods, if not scheduled well, will swap between cache states frequently, thus degrading device performance. The authors propose a framework that combines the thread scheduling strategies with dynamic power mode control strategies together to make a better trade-off between system performance and power consumption. Experimental results show that compared to using only scheduling policies, the systems combining Dynamic Power Management (DPM) and Bank Usage Table (BUT) policies have 14.8% and 9.5% performance growth as well as 9.9% and 13.2% power consumption reduction, respectively. The energy delay product (EDP) is decreased by 20.1% and 22.5%, respectively.

Russia's President Vladimir Putin said: "Artificial Intelligence is the future, not only for Russia but for all humankind. It comes with enormous opportunities, but also threats that are difficult to predict. Whoever becomes the leader in this sphere will become the ruler of the world." So we should have a very convincing argument for its safety before applying an advanced intelligent system. How can we realize that argument is rigorously correct? Famous Computer Scientist Edsger W. Dijkstra said: "The only effective way to raise the confidence level of a program significantly is to give a convincing proof of its correctness." The answer is mathematical proof. Thus, the third chapter, titled "Formal Verification, Testing, Inspection for Intelligent Services," aims to show formal methods, including modeling, specification, verification, and testing techniques, to intelligent services.

The fourth chapter, titled "QoS for 5G Mobile Services Based on Intelligent Multi-access Edge Computing," is about 5G mobile and wireless networks, as well as their cloud computing and QoS mechanisms. Furthermore, a novel advanced QoS concept for 5G mobile services based on Intelligent Multi-Access Edge Computing together with radio network aggregation capability and cloud computing orchestration mechanisms are presented. Besides, network slicing in 5G is also elaborated. Finally, 5G features about vertical multi-homing and multi-streaming for smart end-user terminal devices combined with the capability of radio network aggregation are also clarified. The novelty in the presented concepts and platforms for Intelligent Multi-Access Edge Computing and QoS mechanisms is that they provide the highest level of user access probability ratio, the greatest user throughput, and the greatest number of satisfied smart device users with minimum service cost and optimized utilization of network assets due to the sharing of the traffic load. The performed analysis in this chapter demonstrates that performance gain with the Intelligent Multi-Access Edge Computing module in 5G mobile terminal is higher if there are more available radio access points in comparison with the scenarios with a lower number of radio access points.

The fifth chapter, titled "An Efficient Interpolation Method Through Trends Prediction in Smart Power Grid," discusses that due to the popularity of smart electric-meter, electricity data is fast generated and abundantly transmitted through hierarchical servers in smart power grid domain. The missing record during transmission will influence subsequent analyses. However, it is not trivial to improve the quality of such continuous sensory data, because interpolate accuracy and processing latency are hard to be guaranteed in practice through traditional means. The authors propose a data interpolation method for electricity data of smart meters in a hierarchical edge environment. The missing records would be interpolated by predictive values through support vector regression in the edge environment. In extensive experiments on real data, the accuracy of data interpolation is guaranteed above 90% with the execution time less than 20 milliseconds.

The sixth chapter, titled "2PC+: A High Performance Protocol for Distributed Transactions of Microservice Architecture," introduces microservice-based distributed transaction protocols. In general, distributed storage systems run transactions across machines to ensure serializability in single-service architecture. Traditional protocols for distributed transactions are based on two-phase commit (2PC) or multi-version concurrency control (MVCC). 2PC serializes transactions as soon as they conflict and MVCC resorts to abort, leaving many opportunities for concurrency on the distributed storage system. While in the microservice architecture, service nodes are deployed in heterogeneous distributed systems. 2PC and MVCC are struggling to break through the performance limitations of their existence. The authors present 2PC+, a novel concurrency control protocol for distributed transactions that outperforms 2PC by allowing more concurrency in microservice architecture. 2PC+ is based on the traditional 2PC and MVCC, combined with transaction thread synchronization blocking optimization algorithm SAOLA. Then, they verify the algorithm SAOLA using the temporal logic of action TLA+ language. Finally, they compare 2PC+ to 2PC by applying a case of Ctrip MSECP. When concurrent threads of the service call reach a certain threshold, the RT performance of 2PC+ is improved by 2.87 times to 3.77 times compared with 2PC, and the TPS performance of 2PC+ is 323.7% to 514.4% higher than 2PC.

The Intelligent Service Computing in Mobile Environment is still a hot and emerging research topic. Thus, we wish these chapters can inspire blooming studies on the related topics of service computing, intelligent service, and mobile environment.

Shanghai, China   Honghao Gao
Hangzhou, China   Yuyu Yin

Acknowledgment

We are thankful to our reviewers for their effort in reviewing the chapters. We also thank the Editor-in-Chief, Dr. Imrich Chlamtac, for his supportive guidance during the entire process. We also thank you for the support from Eliska at EAI Office.

Contents

Chapter 1
A Cross-Domain Security Model Based on Internet of Vehicles

Wei Ou, Meiyan Wei, Qin Yi, and Lihong Xiang

1.1 Introduction

The Internet of Vehicles (IoV) refers to the realization with the help of next-generation information and communication technologies to implement all-round network connection between vehicles and vehicles, vehicles and roads, vehicles and persons, and vehicles and service platforms. This technique not only can improve the transportation efficiency and driving experience but also enhances automatic driving ability and the level of intelligent vehicles. Build a new business format for automobiles and transportation services. Provide users with an integrated service which is comfortable, intelligent, energy-saving, safe, and efficient [1]. With "both ends-cloud" as the core, the Internet of Vehicles (IoV) is assisted by roadbed facilities, including intelligent networked vehicles, mobile intelligent terminals, Internet of Vehicles service platform, and other objects. It contains five communication scenarios: vehicle-human communication, vehicle-vehicle communication, vehicle-cloud communication, vehicle-road communication, and intra-vehicle communication [2].

Vehicle network is an ever-changing open network. There are a number of entities like floating cars and various types of driving test equipment in this open network. The most effective way to ensure the security of network is to build a trust mechanism based on the transmission and dissemination of trust. To ensure that the results are more accurate and closer to the real data, we measure and calculate the credibility of the target entity and then choose the data provided by the reliable entity as the processing object.

W. Ou (✉) · M. Wei · Q. Yi · L. Xiang
Hunan University of Science and Engineering, Yongzhou, China

© Springer Nature Switzerland AG 2021   1
H. Gao, Y. Yin (eds.), *Intelligent Mobile Service Computing*, EAI/Springer
Innovations in Communication and Computing,
https://doi.org/10.1007/978-3-030-50184-6_1

1.2 Background

With the continuous expansion of the application scope of Internet of Vehicles, attacks also increase correspondingly. There have been too many attack accidents in Internet of Vehicles. And the intelligent vehicle has become an important target of a hacker. In the recall of Fiat Chrysler automobile company in the United States, hackers used their technologies to break into the "uconnect" system of a running Cherokee Jeep and remotely controlled its acceleration and braking system, radio station, wiper, and other equipment of this car. The BMW digital service system ConnectedDrive has been invaded. Hackers can use the vulnerability to intrude into the vehicle remotely and wirelessly and open the door easily. Considering of Tesla Model S, security experts can open the door and drive away through the loopholes in Model S; at the same time, they can also send a "suicide" command to suddenly shut down the engine when it is running normally. People pursue the convenient, efficient, and pleasant driving experience brought by the IoV, while they also concern or even fear about it.

Security of Internet of Vehicles has been paid more and more attention and becomes a new research hotspot. The UN/WP29 established a special TFSC in 2016, which has formally listed the automobile information security in WP29's work schedule. The working group will carry out related works of automobile information security in three key areas: network security, data protection or personal privacy, and software wireless upgrade. Yifei Wang, CEO of Jidou Internet of Vehicles, said their products use the method of isolating the bottom layer of cars and adding a hardware firewall to ensure vehicles' security. In case of hacker's invasion, the hacker can't get the underlying information of the vehicle and cannot control the vehicle. And the relevant data of the vehicle can't be obtained and tampered with. At the software and cloud level, Jidou has also done lots of works and formed a comprehensive set of protection. However, most of these networking devices are connected by means of OBD interface and other forms. If hackers want to attack, they need to ensure that the OBD device is in vehicle and they cannot be far away from the vehicle. So the possibility of attacks is little.

Based on the comprehensive analysis of security incidents of IoV in recent years, there are three major risks [3]: ① The vehicle networking architecture is vulnerable to challenges of information security. ② Wireless communications is facing more complex communication environment. ③ There are more potential attack interfaces in security management of the cloud platform.

In this paper we focus on two technologies: the cross-domain security technologies and the trust evaluation method. In the security model of IoV, cross-domain technologies are used to realize transactions among different domains of IoV. The method of trust evaluation is used to ensure transactions security and construct a trustworthy environment for cross-domain of IoV.

1.3 Cross-Domain Security Model of IoV

1.3.1 Division of Security Domain

In order to meet needs of safety management, according to the vehicle networking architecture [4] and safety level and safety requirements, we divide the security domain into three layers: cross-domain application layer, cross-domain network layer, and cross-domain perceptual layer. By defining three security domains, we can effectively implement definitions of security scopes to centralize management of security threats and effectively improve the efficiency of transactions between two and more chains of IoV.

The reason that the security domain is divided in this paper is that the security domain has the three following advantages: (1) Security threats faced by the same security domain are similar, which can facilitate the classification of cross-domain security and effectively solve problems of cross-domain security. (2) The security domain mainly uses trusted computing security technologies to ensure the credibility of transaction chains, but different security domains have different technologies. Combined with technologies of various security domains, it is possible to achieve effective cross-domain security services. (3) In transactions of cross-domain security model, information is input and exchanged at the beginning of the transaction, and then it is processed in the security domain [5].

Figure 1.1 shows the division and protection system of cross-domain vehicle networking security domain.

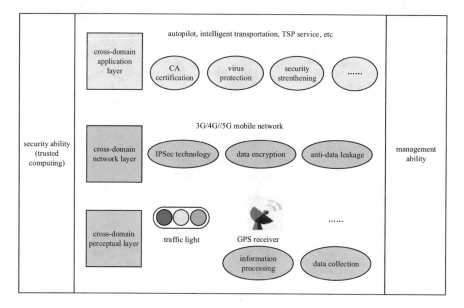

Fig. 1.1 Division and protection of security domain

1.3.2   Security Model

In order to ensure the credibility and security of IoV, we propose a cross-domain security model based on IoV. In our model the secure domain is divided into three layers, which includes the cross-domain application layer, the cross-domain network layer, and the cross-domain perception layer. The method of trust evaluation is used as the main technology of secure protection.

1.3.2.1   Model Structure

Communication characteristics of IoV restrict its security and communication ability [6]. Security ability of this model includes technologies such as trusted computing, which provides key management and identity authentication. It ensures the authenticity of vehicles' identity information in network, provides information protection, and ensures that the transmission data will not be damaged or distorted.

Cross-domain security architecture of Internet of Vehicles is shown in Fig. 1.2.

1.3.2.2   Protection Strategies of Security Domain

In cross-domain vehicle network, different security domains are facing different threats. In our paper, security problems and corresponding countermeasures are analyzed in three security domains. The three security domains include cross-domain application layer, cross-domain network layer, and cross-domain perception layer [7].

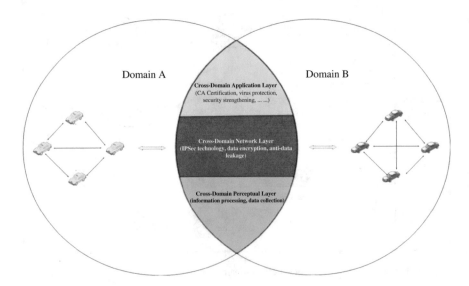

Fig. 1.2 Security architecture of Internet of Vehicles

1. Security Protection of Cross-Domain Perception Layer

In cross-domain vehicle networking, the perception layer undertakes the overall perception and collection of information of the onboard unit and the base station. When perceptual information is transferred to the transport layer, it is necessary to ensure the authenticity and effectiveness and availability. Base station [8] provides communication links for servers and terminals to complete interactions of data and control information between them. In general, the automatic identification technology RFID is used in the perception layer of cross-domain vehicle network. While in condition of wireless transmission, attackers can easily intercept or tamper with sensitive information when signals are transmitting between the nodes. Therefore, the protection design of cross-domain perception layer includes two parts: data collection and information processing. It is mainly guaranteed by designing a security protocol [9] which is suitable for the cross-domain perception layer of IoV. The specific process is explained as follows:

① The built-in reader in the base station continuously sends radio frequency signals covering a certain range. When a vehicle is found entering its RF range, it will send an authentication request BSHello and wait for this vehicle to receive it.

② Vehicles entering the working range of the base station will receive the message BSHello from the base station, including two random numbers R1 and R2, as well as the hash algorithms supported by the base station, which are selected for the vehicle units.

③ The onboard unit returns the message TagHello, which contains the hash algorithm selected by the onboard unit and the TagID of the electronic tag of the OBU.

④ After the base station certifies the validity of the vehicle unit, the session key for communications between two parties is calculated by a reasonable algorithm.

⑤ The base station sends the message BSResponse to inform the vehicle unit of the newly generated session key, which contains the initial key of the electronic tag of the OBU. It is used to verify the validity of the base station.

⑥ Application data are interactive between the base station and the vehicle unit, all of which use the agreed session key to encrypt application data to ensure information security.

⑦ When the onboard unit drives out of one base station range and enters another base station range, it will connect with the new base station by the same authentication way, discard the original session key, and record the new session key.

In Fig. 1.3 it shows the process of secure communication protocol between the OBU and the base station.

2. Security Protection of Cross-Domain Network Layer

Deployments of firewalls and intrusion detection systems can effectively prevent and detect kinds of network attacks, realize secure transmissions of vehicles' electronic information collected by front ends of IoV, and ensure the integrity

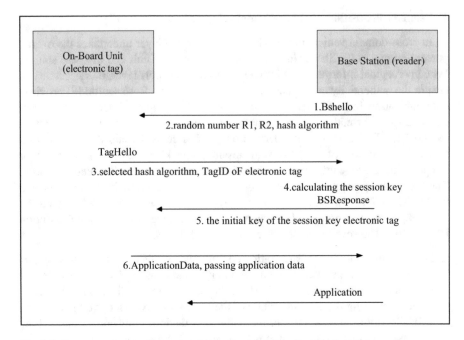

Fig. 1.3 Secure communication protocol between onboard unit and base station

and consistency of data. It is necessary to set the data center of the cross-domain vehicle networking as the center, construct the tunnel encryption system IPSec, and transmit the encrypted data collected by base stations of the cross-domain vehicle networking, so as to prevent the data illegally tampered and intercepted. In addition, for base stations transmitting information by way of wireless, the encrypted wireless tunnel should be used to upload the wireless basic information [10].

For internal security of vehicle networking [11, 12], data encryption technologies are generally used to solve problems of data confidentiality leakage. For the problem of camouflage and forgery, verifications of authenticity and legality of communication entities and data are required. For the problem of entities being tampered, effective methods are needed to protect the entity integrity. In the case of denial of service attacks, if attacks occur from outside, the attack behaviors can be detected and filtered by firewalls, gateways, etc. And the attack behaviors will be blocked out of subnet. If attacks occur in subnet, effective methods are needed to detect the attack behaviors so as to provide alarm information for users of IoV. The Internal security of IoV is directly related to the security of personal and property. A single means of protection can only meet partial needs. So it is necessary to combine a variety of different security strategies.

In Fig. 1.4 it shows the network security components in vehicles.

① Security Certification: Firstly, entities in vehicle networking are certified to protect the integrity of each ECU (electronic control unit) node. Secondly, mutual

Fig. 1.4 Secure component
in vehicles

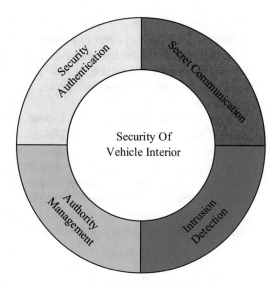

certifications between each ECU node verify the compliance of each ECU node participating in communications. Based on these, the authenticity and integrity of the data in process of vehicle networking communications are verified.

② Confidential Communication: In process of communications of IoV, the data are encrypted to ensure the confidentiality.

③ Authority Management: During the design and implementation of vehicle networking, the access authority of each area should be managed and controlled. At the same time, it is necessary to filter the interactive data between subnets and control access authority of the CAN bus of the onboard entertainment system and other equipment. Access control and authority management are very important for system security.

④ Intrusion Detection: As a supplement, in addition to the passive defense of attacks, the intrusion detection mechanism should be more active. For the CAN bus of vehicle networking, by detecting whether there are illegal messages on the CAN bus, attack behaviors will be detected.

3. **Security Protection of Cross-Domain Application Layer**

　① It is difficult to prevent and detect by firewalls and intrusion detections when internal users abuse network resources and use open services for unauthorized accesses, illegal operations, or unintentional damages. Therefore, it is necessary to deploy an audit system in the business application area and the storage backup area, so as to monitor and audit the access behaviors of all users.

　② In condition of complex functions and huge codes, there are some security vulnerabilities and unknown "back-doors" in operating systems, database systems, application software systems, and some equipment systems, which are usually found. So it is necessary to deploy a vulnerability scanning system in the business application area and storage backup area of the data center to detect services regularly. It is important to check configurations of operating systems and database systems of devices, potential secure hazards, and security risks. It is helpful for the security administrator to control possible safety events and eliminate potential secure hazards as much as possible. At the same time, the terminal patch distribution function and manual reinforcement should be performed.

③ In order to prevent the host system from being attacked by external or internal viruses, malicious codes, Trojans, etc., it is recommended to deploy antivirus system in network management area of the data center and install antivirus client programs on the terminal and server.

④ In order to protect the open application services of the operation data center, it needs to deploy another layer of application firewall on the basis of the external firewall protection to realize the security protection of the business application layer and avoid the system from various application attacks.

1.3.3 Construction of Inter-Domain Trust Relationship

1. **Calculation of Inter-Domain Recommendation Trust**

In security domains, inter-domain recommendation trust [13] refers to the recommendation of trust between trust agents that two nodes from different domains judge the trust of each other. If there is a direct transaction between the neighbor node and the target node, then the recommendation trust can be calculated. Conversely, if there are no direct transactions between neighboring node and target node (such as service providers), a recommendation trust path must be found by the domain agent. Here, the inter-domain trust relationship can be abstracted into a directed graph. Each node in the graph denotes a domain. The edges of the graph denote the trust relationship between each domain. Recorded as the directed graph, $G = (V, e)$. The node (service applicant) needs to send a trust recommendation request to the domain agent and reports the basic information of the target node to the trust agent. After that, the trust agent has two things to do. One is to find the optimal path, and the other is to calculate the trust value of the target node. As shown in Fig. 1.5.

Describe the situation above as $G = (V, E)$. We use the shortest path maximum trust method to choose the optimal path. Shown in Fig. 1.5, there are three paths: ta2-ta3-ta4, ta5-ta6, and ta7-ta8. First, choose the shortest path. We can see that ta5-ta6 and ta7-ta8 are the shortest paths. We then choose the maximum value of the trust from both paths. The calculation method of trust value is as below:

$$\text{rtv}_{B_j}^{A_i} = \text{dt}_{B_j}^{TA} * \prod_{k=m}^{n-1} \text{rtv}_{TA_{k+1}}^{TA_k} \tag{1.1}$$

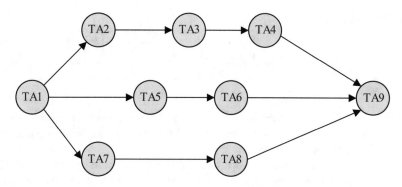

Fig. 1.5 Recommendation inter-domain trust

2. **Updating of Inter-Domain Trust**

If transactions between two nodes from different domains are successful, the recommended inter-domain trust value will increase; otherwise it will decrease. Considering of the specific increment or decrement, the number of successful or failed transactions between the nodes Ai and Bj should be the first priority.

$$\begin{cases} \text{rtv}_B^A = H + \mu \times \varphi(s) & Transaction\ Successed \\ \text{rtv}_B^A = H - \mu \times \varphi(f) & Transaction\ Failed \\ \varphi(x) = e^{-1/x} \end{cases} \quad (1.2)$$

3. **Trust Maintenance**

We manage the trust value of each node by setting the trust management unit. When the trust value of one node saved in unit changes, the management unit will update its trust value by the corresponding algorithm. After obtaining the direct trust, according to the calculation method of recommendation trust and overall trust, the corresponding values in the table can be obtained, and then generation and updating of the whole trust relationship table can be realized.

① **Construct Trust Relationship Table**

In each cluster, a table is designed for the fusion node to save trust values of all nodes in the cluster. According to the trust relationship between the fusion node and the source node, the corresponding trust value in this trust relationship table is determined. In addition, total trust in the table is the result of combining direct trust and recommendation trust.

In order to evaluate the trust value of a certain node, the fusion node will access and view the trust relationship table. If the trust value of a node is less than the setting threshold value, it means that this node is not trustworthy. And related communication will be terminated. Corresponding rules will be used to

Table 1.1 Trust relationship

Node	1	2	...	j
Direct trust	$DT_{p1}(t)$	$DT_{pj}(t)$	...	$DT_{pj}(t)$
Recommendation trust	$RT_{p1}(t)$	$RT_{pj}(t)$	...	$RT_{pj}(t)$
Total trust	$CT_{p1}(t)$	$CT_{pj}(t)$	...	$CT_{pj}(t)$

punish it, and the punishment record will be generated. If communications cannot be terminated, some effective evaluation methods should be referred, and related communications must be cautious.

Because of the time decay of trust, influences of time should be considered. That is, historical high trust value cannot reflect current trust of the node. Therefore, at a certain time t, the trust value obtained by the source node i from the fusion node p is expressed in the form of $(\cdot)T_{pi}(t)$ (Table 1.1).

In order to show that the fusion node p and the source node i do not have any historical interaction behaviors, the trust relationship table is initialized, and values in the table are set to the fixed values. At certain time t, when the fusion node p evaluates the source node i, the direct trust value is calculated according to the following formula:

$$DT_{pi}(t) = (1 - \alpha) DT_{pi}(t_0) \times \beta(t - t_0) + \alpha \times LT_{pi}(t) \qquad (1.3)$$

In formula (1.3), $(0 \leq \alpha \leq 1)$ represents weighting factor; $\beta(t - t_0)$ is the time decay factor. Considering the complexity of reality, the concept of trust level is introduced to evaluate the trust value. The trust level is defined as $LT_{pi}(t)$, which expresses the trust level of the fusion node p to the source node i at time t. It can be seen from the above formula that the effect of $LT_{pi}(t)$ on $DT_{pi}(t)$ can be adjusted by α. In Internet of Vehicles, fusion nodes use method of data analysis or system detection to evaluate specific behaviors of the source node i in a period of time $t - t_0$ and obtain the fixed parameter $LT_{pi}(t)$. At the same time, the detection system records all malicious behaviors of node i.

② **Updating of Trust Value**

The trust relationship table is obtained by calculations of trust values. In order to update values in the table in real time, the corresponding updating algorithm is proposed. During the first data fusion, the initial trust values of all nodes are set to 1. When the data fusion is finished, the new calculated value is used to replace current trust value in the table.

Firstly, the direct trust value of the source node i at time t is obtained according to the formula. Secondly, by analyzing the value, the node i is judged whether it is a malicious node and decided whether to continue to communicate. Finally, the evaluation results are counted, and the trust value of the source node is updated. We record the success and failure times of communications between the source node and the fusion node to evaluate the statistics. According to the results, we construct the following functions to complete the updating of trust value:

$$DT_i(t) = \begin{cases} DT_i(t) + \varepsilon_1 e^{-1/x} & 0 < \varepsilon_1 < 1 \quad (F = 1) \\ DT_i(t) - \varepsilon_2 e^{-1/x} & 0 < \varepsilon_2 < 1 \quad (F = 1) \end{cases} \tag{1.4}$$

In formula (1.4), ε_1 and ε_2 are the updating coefficient, x is the number of contacts, and F is the symbol of success or failure. When value of F is 1, it means communications are successful, and when value of F is 0, it means unsuccessful communications. It can be seen from the above formula that when communications are successful, direct trust values of nodes $DT_i(t)$ will increase appropriately, while failure communications will cause the value to decrease appropriately.

Assuming ε_1 and ε_2 are fixed values. Times of success or failure communications directly determine increase or decrease of trust value. The larger value x, the faster the change of trust. From the subjective aspect, compared with success factors, failure factors have greater impact on the trust evaluation. So malicious nodes can be punished by setting the value of ε_2. When the direct trust value of the source node is less than the setting threshold, the trust value of the node is reduced by a large increase of ε_2; thus behaviors of malicious nodes are punished.

In summary, considering of actual communications, adjusting the update coefficient, and rewarding and punishing according to behaviors of nodes, finally the real-time updating of the trust relationship table is finished.

4. **Evaluation Process**

Firstly, according to similarity characteristics of information collected by adjacent nodes in vehicle networking, we perform the preliminary screening of trusted nodes. Secondly, we calculate trust values of all sensing nodes and complete the trust evaluation by grouping and fusing all nodes according to trust values.

When the source node uploads data to one node, this node needs to perform the following operations on the source node:

① Judge whether this node is an in-group node. Analyze the physical location of the source node. If the location of this node is similar or adjacent, it is an in-group node. According to the evaluation algorithm of in-group nodes, the node is preliminarily screened. If characteristics of this node are not met, further analysis will be carried out.

② Search the direct trust table. If information of this node exists, turn to ④. If not, the analysis will be continued.

③ Query the recommendation trust relationship table. If information of this node exists, turn to ④. If not, the uploaded data will be discarded, and the failure access will be recorded.

④ Trust evaluation. Calculate trust values of nodes. If the trust value is higher than the setting threshold, accept the uploaded data and turn to ⑤. Otherwise, discard the data.

⑤ Update the trust relationship table.

⑥ Remove the node from chain if the node's trust value is under the setting threshold.

⑦ Finish the evaluation and summarize all accesses.

The trust evaluation process of IoV is shown in Fig. 1.6.

1.3.4 Construction of CA Infrastructure

Framework of security infrastructure is shown in Fig. 1.7. In Fig. 1.7, the manufacturing factory is responsible for the production of related equipment of cross-domain vehicle networking, such as terminal equipment, roadside equipment, secure equipment in background system, etc. During the production process, the unique identification of the equipment will be written, which will not be changed in the whole life cycle of the equipment. In security production environment, the original authentication and authorization mechanism are written into the equipment, through which the default information of credit registration institution and credit authorization institution can be written into the equipment.

The registration authority is responsible for the certification of onboard equipment and roadside equipment. Only after the certification of relevant registration authority, the equipment can be used in the system. The registration authority first verifies whether the device is legal and then issues the certificate for the legal device, that is, the registration certificate. Authentication certificate is used to apply for terminal authorization certificate.

The authorized organization is responsible for the authorization of onboard equipment and roadside equipment. Only with the authorization of relevant authorized organization, these equipment can broadcast or receive the message of authorization permission in system. The authority first verifies the validity of the authentication certificate issued to the device and then issues the authorization certificate for the legitimate device, that is, the security message certificate or the service message certificate.

1.3.5 Basic Workflows of CA Management System of Model

1. The registration organization authenticates qualifications of equipment of IoV and service organizations and issues registration (authentication) certificates to authenticated entities
2. The equipment of cross-domain vehicle networking applies for the authorized functions of IoV to the authorized institution by the registration certificate.
3. The authorized organization issues the authorization certificate to the equipment of IoV according to its authentication certificate. Functions and safety operations that the equipment can perform are described in the authorization certificate.

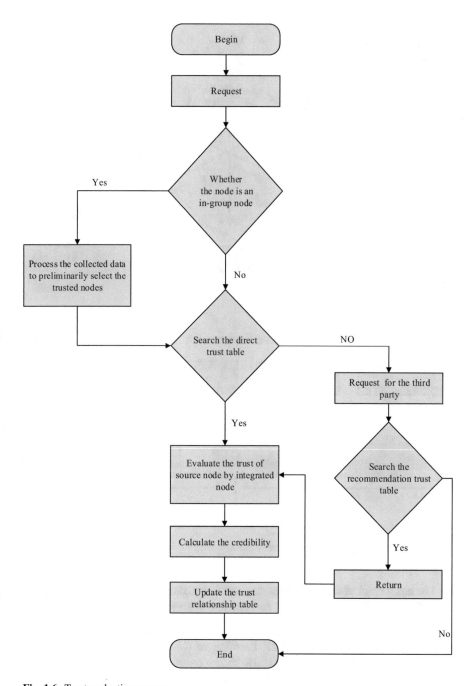

Fig. 1.6 Trust evaluation process

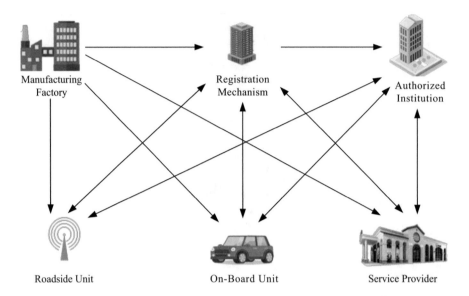

Fig. 1.7 Framework of security infrastructure

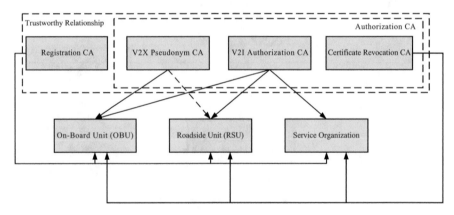

Fig. 1.8 Deployment of distributed security infrastructure

4. The equipment of IoV uses its authorization certificate and corresponding public and private keys to sign, verify, encrypt, and decrypt.

Aimed at active secure businesses for vehicles, we propose a PKI system consisting of Registration CA, V2X Pseudonym CA, V2I Authorization CA, and Certificate Revocation CA, realize active secure businesses for vehicles, and protect users' privacy in cross-domain vehicle networking. In Fig. 1.8 it shows a possible CA deployment scheme Distributed Security Infrastructure Deployment, which realizes a trusted PKI system by cross-certifications among CAs.

Functions of parts are as follows:

1. Registration CA. Responsible for issuing registration certificates to vehicle terminals, roadside units, service organizations, and other entities, which meet the conditions of network access. These entities use registration certificates to further apply to other authorized CA for certificates to achieve certain secure communication capabilities.
2. V2X Pseudonym CA. Responsible for issuing pseudonym certificates to vehicle terminals for anonymous communications in vehicle to vehicle. The certificates are used to sign and issue basic safety messages (BSM) to protect users' vehicle identity anonymously.
3. V2I Authorization CA. Responsible for issuing certificates to vehicle terminals, roadside units, and service organizations for secure communications between vehicles and roadside units.
4. Certificate Revocation CA. Responsible for issuing revocation lists of various certificates.

This distributed deployment scheme can be used to set different root CAs for different businesses. But it needs to construct trust relationships between different root CAs. It can be applied to the scenario that multiple parties jointly manage and maintain the vehicles in cross-domain of Internet of Vehicles. The schemes' advantage is that it is easy to be connected to the existing management mechanism and the corresponding functions can be added to the existing CA system.

1.4 Conclusion

In this paper we discuss related works on Internet of Vehicles. Based on these, we look forward to constructing a cross-domain security model of Internet of Vehicles. Firstly we introduce the security domain and analyze its divisions. Secondly we study the structure of cross-domain of Internet of Vehicles. Finally we divide the security domain of Internet of Vehicles into three layers which are cross-domain application layer, cross-domain network layer, and cross-domain perception layer. Also we analyze possible threats and the corresponding strategies in the three security domains and propose a cross-domain security model of Internet of Vehicles. At the same time, we study the trust relationship in IoV. The method of trust evaluation is used to ensure transactions security and construct a trustworthy environment for cross-domain.

In the next step, we will work closely with automobile manufacturers, pay more attention to industrial demands from automobile manufacturers, and seek more efficient and convenient security solutions. At the same time, we will build a test platform to promote the test verification and optimization of technical scheme in stages and provide test basis for the implementation of secure technical scheme.

Acknowledgments This project is completed under the support of the construct program of the applied characteristic discipline of Hunan University of Science and Engineering.

References

1. *Security of Internet of Vehicles [EB/OL]* (China Information and Communication Research Institute, 2017)
2. C. Shen, *Research on Communication Security and Privacy Protection in Internet of Vehicles [D]* (Beijing Jiaotong University, 2018)
3. C. Xu, Internet of vehicles information security threats and protection strategies [J]. Inf. Commun. **07**, 191–192 (2018)
4. R. Liu, *Research on Information Security and Privacy Protection Mechanism of Vehicle Network [D]* (University of Electronic Science and Technology of China, 2018)
5. Y. Jiang, *Quantitative Evaluated Model Based Security Domain between Network Security Strength and Network Delay [D]* (Hunan University of Technology, 2016)
6. W. Du, J. Deng, Y.S. Han, et al., A pairwise key pre-distribution scheme for wireless sensor networks [J]. J. ACM Trans. Inf. Syst. Secur. **8**(2), 228–258 (2005)
7. X. Wang, *Research on Model of Creditability-Based Data Fusion for Internet of Vehicles [D]* (Chang' an University, 2014)
8. Z. Yang, *Research on Security Mechanism and Key Technologies in Vehicular Networks [D]* (Beijing University of Posts and Telecommunications, 2019)
9. L. Bu, *Research on Safety Architecture of Expressway Internet of Vehicles System [D]* (Tianjin University, 2012)
10. N. Chen, Design and analysis of internet of vehicles safety protection system [J]. Comput. Dev. Appl. **27**(10), 32–34+37 (2014)
11. C. Wu. *Research on Key Technologies of Vehicle Internal Network Security for Internet of Vehicles [D]* (Southeast University, 2018)
12. X.L. Liu, *Research on OBU-based Multilevel Security Architecture and Communication Scheme for Internet of Vehicles [D]* (Jiangsu University, 2018)
13. Z. Zhang, *Study on Grid Multidimensional Trust Model Based on Fuzzy Comprehensive Evaluation [D]* (Qufu Normal University, 2014)

Chapter 2
A Framework to Improve Performance and Energy Efficiency of Embedded Intelligence Service Systems

Yikun Xiong, Zongwei Zhu, Jing Cao, Junneng Zhang, Fan Wu, and Huang He Liu

2.1 Introduction

The computing capability is the key determinant of the performance of embedded services. However, the increasing demand for intelligent services brings higher requirements to intelligent service systems. Most *artificial intelligence* (AI) applications have large data volume, complex data structure, and high computational density, which brings severe challenges to the computing capability of embedded intelligent service systems. What's worse, according to researches [7, 13], the data volume is still proliferating. Furthermore, most terminal devices of embedded intelligent service systems, such as edge computing systems, have high real-time requirements but limited resources, such as the storage space, the number of processors, and the battery capacity. As a result, the need to increase the processor performance of the terminal devices while reducing power consumption and storage size is imminent.

The usual means to improve the processor performance is to increase the frequency and the number of cores. However, they all bring higher power consumption, which is not feasible for embedded intelligent service systems. Therefore, the academia and industry turn to explore multi-core technology. Usually, there are shared secondary or tertiary caches between multiple cores to store the most commonly used data. With the development of multi-core architectures, the increasing number of threads will inevitably lead to fierce cache competition during job scheduling [2, 11]. So far, there have been many studies to alleviate this problem. The thread group scheduling strategy is a widely used multi-core processor

Y. Xiong · Z. Zhu (✉) · J. Cao · J. Zhang · F. Wu · H. H. Liu
Suzhou Institute for Advanced Study, University of Science and Technology of China, SuZhou, China
e-mail: SAxyk999@mail.ustc.edu.cn

© Springer Nature Switzerland AG 2021   17
H. Gao, Y. Yin (eds.), *Intelligent Mobile Service Computing*, EAI/Springer Innovations in Communication and Computing,
https://doi.org/10.1007/978-3-030-50184-6_2

cache management method and has been studied a lot. *Tam* et al. [15] choose and schedule parallel running threads that compete the least in last level sharing cache. They reasoned that if the cache is highly shared, the threads will have highly degree of parallelism between them. The scheduler works well in solving the cache competition problem according to their experimental result. In the realm of databases, *Harizopoulos* and *Ailamaki* [11] try to build an effective thread context switching strategy for improving the reusability of instruction cache (I-Cache). *Snavely* and *Tullsen* [2] demonstrate the regular scheduling is defective; based on their experimental platform, they proved that the symbiotic thread group scheduling is feasible in performance-enhancing. However, they all perform poorly in power optimization.

The energy consumption issue has also received much attention. The dynamic power state management strategy mentioned in [8] introduces a certain threshold. When the idle time exceeds the threshold, the power mode of the chip is dynamically converted during the same control epoch. The BUT strategy mentioned in [16] uses a *Bank Usage Table* (BUT) to mark the usage of the *Rank* and adjust the power mode of the *Rank* based on the flag. However, if not scheduled well, the frequent swap in and out of cache states will bring unnecessarily conversions of processor's power mode, thus degrading the performance of the device.

In this paper, we propose a framework to balance the performance and power consumption of embedded intelligence service devices. The framework adapts *Application-Based Threads Group Scheduler* (AGS) [18] to achieve thread aggregation and combines it with the dynamic power mode control strategy together. The framework includes the *Information-Collecting layer*, the *Decision-Making layer*, the *Interface layer*, and the *Power Mode Adjustment Module*. The system collects statistics on the operating system (OS) resource usage and transmits them through the information exchange component. This information is used by the power management strategy to formulate energy optimization method. To facilitate our subsequent research on selecting the optimal strategy in multiple strategies, the generated method is then executed by the *Decision-Making* component in the *Decision-Making layer*. In order to further improve energy efficiency, we also introduce a free pages allocator called *RS-buddy System* to centralize the allocation of memory. The *RS-buddy System* can gather the memory usage into as few *Ranks* as possible, thus improving the idleness of the *Rank*. The concentrated memory usage makes as much memory as possible in the low power mode, thus achieving better effect of reducing power consumption.

Experimental results show that when we combine AGS with DPM, the *Rank Idle Rate* is reduced by 5.9%, the power consumption is reduced by 9.9%, and the EDP is reduced by 20.1%. When the AGS and BUT strategies are combined, the *Rank Idle Rate* is reduced by 11.3%, the power consumption is reduced by 13.2%, and the EDP is reduced by 22.5%.

The rest of the paper is structured as follows: the Sect. 2.2 describes the related work which our research is based on. We will present our proposed framework in detail in Sect. 2.3. The experimental verification is in Sect. 2.4. In Sect. 2.5, we conclude the paper.

2.2   Related Work

Because of the various limitations of single-core processor technology development, many researchers are turning their research focus to multi-core architecture. Multi-core architecture allows multiple processes to run independently in parallel. However, when cache resources shared between cores are used, different processes will mutually interference with each other, thus increasing the execution time [3]. As a result, the device performance can be reduced largely as the cache competition intensifies. In order to solve this problem, many strategies have been proposed. The contention of the cache space is not all factors that cause the performance depreciation; *Sergey Zhuravlev et al.* [12] find that the memory bus, controller, and prefetching hardware contention are also contributing factors.

In order to mitigate the effects of these factors, they develop scheduling algorithms DI and DIO to allocate threads, so as to ensure that the cache miss ratio is uniformly distributed among the caches. The thread grouping scheduler proposed by *Tam* et al. [15] chooses and schedules parallel running threads that compete the least in the last level of sharing cache. They reasoned that highly sharing cache mean high parallelism among threads. The scheduler works well in solving the cache competition problem according to their experimental result. *Snavely* and *Tullsen* [2] demonstrate regular scheduling is defective; on their experimental platform, the symbiotic thread group scheduling is proved to improve performance. There are also many performance optimization strategies at other software level. For example, in the realm of databases, *Harizopoulos* and *Ailamaki* [11] try an effective thread context switching strategy for improving the reusability of instruction cache (I-Cache). In terms of hardware, *Bellosa* and *Steckermeier* [4] are the pioneers who detected the sharing among threads by means of the hardware performance unit (PMU). The PMU is then used by many researchers to achieve better performance optimization. *Parekh* et al. [14] use the cache miss information provided by PMU to perform intelligent thread group scheduling. *Weissman* [17] suggested to use the PMU to detect the cache misses and reduce the cache trashing on account of cache capacity limit (capacity misses) and data replacing due to the address conflicts (conflicts misses), but the shared regions among threads needs to be manually identified. *El-Moursy* et al. [1] and *McGregor* et al. [9] find that it is not feasible to determine the best collaborative schedules only deal with cache interference on multi-processors. *Philbin* et al. [5] attempts an automated parallelization method to increase the cache sharing reuse for a single-threaded sequential program. These previous studies mainly utilize the characteristics of resource sharing and the similarity of cache states during the scheduling between different threads, and they can make threads run more efficiently. If the resource utilization becomes more centralized and efficient, more idle resources will be available. In such a case, the power consumption of the idle resources could be reduced by introducing a power management strategy.

In the mean time, some strategies for power mode management have also been proposed. Memory is regarded as one of the main consumers of power

[6], and memory devices with multiple power modes have been developed by memory manufacturers, which has also become the basis for subsequent power mode management strategies. For the purpose of reducing power consumption better and saving more energy, many hardware-based and software-based power mode management strategies have been proposed. *V. Delaluz* et al. [16] first experimentally used the task scheduler to intervene in the memory power modes management. The OS scheduler tracks the Rank actually used by each thread and records it in a Bank Usage Table (BUT). When one thread is scheduled, the Rank belong to it needs to run in normal power mode; correspondingly, other Rank would be allowed to transit to low power mode. *Lebeck* et al. [8] first uses the sequential first-touch policy to aggregate pages and integrate with DPM strategy to control the memory power mode.

However, performance optimization strategies mentioned earlier focus only on solving cache competition and resource reuse problems, and they all perform poorly in power optimization. The power optimization strategies, however, if not scheduled well, will swap in and out of cache states frequently, thus bringing unnecessarily conversions of processor's power mode and degrading the performance of the device. To better cater to the modern intelligent services on edge devices, both the system performance and energy efficiency should be taken into consideration. However, the combination of the two methods has not been completely and systematically proposed. As a result, in this paper, we propose a framework to balance the performance and power consumption of embedded intelligence service devices. In the next section, we will describe the internal implementation mechanisms of the framework.

2.3 Proposed Framework Mechanism

In this section, we describe the overall architecture, the workflow, and implementation principles of the framework in detail.

2.3.1 The Overall Architecture

To offer better intelligent service on embedded devices, this paper aims to improve the system performance while degrading the power consumption. We propose a hierarchical architecture framework that facilitates substitution between different strategies. The framework includes the *Power Mode Adjustment Module*, the *Decision-Making layer*, the *Information-Collecting layer*, and the *Interface layer*. The *Power Mode Adjustment Module* periodically sends an information request to the *Interface layer*. After receiving the response, it generates the power adjustment method according to the obtained resource utilization information. The method is performed at the *Decision-Making layer*, which is designed to facilitate our

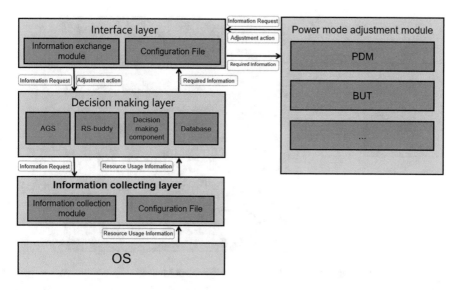

Fig. 2.1 The overall architecture of the framework

subsequent study to select the optimal strategy in multiple strategies. The *Decision-Making layer* contains an application-based thread group scheduler named AGS for centralized scheduling of threads, free pages allocator to allocate memory centrally called *RS-buddy System*, and a *Decision-Making* component for performing power adjustment actions, it is also responsible for sending and receiving information requests. When an information request is received from the *Interface layer*, the *Decision-Making* component will pull the required information from the Information-Collecting *layer* and returns it to the *Interface layer* and performs these actions after obtaining the power adjustment actions from *Interface layer*. The Information-Collecting *layer* is used to collect the utilization information of memory. The last updated data is stored in the configuration file, which is to facilitate to pull data by *Decision-Making layer*. The *Interface layer* is used to dock the *Power Mode Adjustment Module* and the *Decision-Making layer*. It is responsible for receiving information requests and returning information and the variable conversion. Figure 2.1 shows the overall workflow of the framework. Below we'll cover each *layer* in detail.

2.3.2 *Power Mode Adjustment Module*

The *Power Mode Adjustment Module* has two main responsibilities: generate the power management policy and optimize it according to system dynamic characteristics. It can also be replaced with other strategies to achieve other optimization

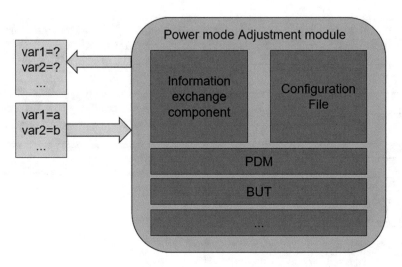

Fig. 2.2 Information processing of Power Mode Adjustment Module

goals. In this paper, we use the DPM strategy [8] and the BUT strategy [16] as our optional power mode optimization strategy.

The DPM strategy can be divided into static and dynamic adjustment strategies. In static policies, all chips return the same basic power state if there is no pending access [8]. For better performance and energy efficiency, the restriction has been relaxed. That is to say, in dynamic adjustment policies, each chip is allowed to reside in different power mode based on their individual access pattern. The DPM strategy specifies a threshold. If the time is beyond the threshold and the chip is still not accessed, the chip can go to the next power mode during one control epoch. The time between accesses to chip is the metric of transition to lower power states.

BUT is a scheduler-based power mode management strategy that manages the *Rank* state according to its state of use. It uses a *Bank Usage Table* to store the information of *Rank* used by the process. *Rank*s in the BUT that need to remain active are flagged, while the unmarked *Rank*, such as idle *Rank*, will then be adjusted to low power state.

In this paper, we introduce the above two power adjustment strategies in the *Power Mode Adjustment Module*. The information exchange component is used to send the information request and receive the information from the *Interface layer*. And it is also responsible for the conversion or mapping of variables when the system and power management strategies use different variables. The module also temporarily stores the received information to the configuration file. Figure 2.2 shows the processing flow of the *Power Mode Adjustment Module*.

2.3.3   Decision-Making Layer

The *Decision-Making layer* includes *Decision-Making* component, *RS-buddy*, AGS and a database. The database is used to store information temporarily. AGS is used to group threads to achieve central scheduling, and *RS-buddy System* is used to centrally allocate memory. In our framework, we use AGS as our default scheduling strategy instead of the Linux default *Completely Fair Scheduler* (CFS) policy. Linux CFS takes the CPU usage time as a key and stores the scheduled tasks records in a *red-black tree*, and the leftmost record in the tree has the least CPU usage time. Each CPU's tree can be used to determine the task with the least CPU usage time. After the thread is scheduled, its processor usage time will be recalculated and then the task reinserted into the tree. In other words, the threads are organized into *red-black tree* according to the weight of each thread. The CFS policy selects the leftmost thread from rb_tree to execute each time. After the thread is selected, the thread will be deleted from the rb_tree. After that, according to the calculation formula of CFS, the position of the thread in the red-black tree would be redetermined. However, the CFS scheduling policy may cause frequent switching of the cache state, which can lead to increased cache competition. Therefore, in this paper, we choose the AGS strategy, which makes full use of the similarity between threads, so as to achieve better results. The mechanisms we use are described in detail below.

2.3.3.1   Application-Based Threads Group Scheduler

In the *Decision-Making layer*, the *Application-Based Threads Group Scheduler* (AGS) we previously proposed is used. The AGS policy is based on several features of the Android programming model:

Highly Shared Memory Address Space In Android programming model, when an application's component starts, Android will start a new process, which contains only one execution thread named main thread. Then, all threads of the application are forked from the main thread. These forked threads from the same main thread share the same memory address space and communicate frequently among them.

Highly Interleaved Scheduling Android is a Linux-based multitask operating system. After the application is closed, the background thread belonging to the application continues running. However, there are also some other threads running in the background to be scheduled, thus causing multiple threads to be interleaved. The cache state of the application running background may be swapped out, thus interfering with the normal use of the application.

Single Application Monopolize the Whole Screen Window Android runs the application in a single window, which means that the currently running application monopolizes the entire screen and hides the existence of threads running in the background from the user. As a result, a bad thread scheduling may slow down the response time of the currently running application.

In AGS, the current application thread is organized by a cycle list. Each thread group has a group number. When performing thread scheduling, the scheduler first checks threads in current thread group. If no thread can be scheduled, it then sequentially searched for the next thread through loop list. The scheduled thread will be marked, and each thread should only be scheduled once in the same scheduling round. After all the threads in the thread group are scheduled, the CFS will be used to find the next thread to schedule. Then, the thread group where this thread is located is marked as current thread group. The scheduler then loops the above process until all the scheduling work finishes. For multi-core processor, each core uses a *cur_id* domain to mark the current thread group. For instance, if there is a quad-core processor sharing the last level of cache, the core i uses *cur_uid*[i] as a domain to mark the current thread group. Thread scheduling on quad-core processor is performed in following manner: if threads in the same thread group on core 1 are all marked as 1, the scheduler will select the thread with the least CPU usage time in the red-black tree. When all threads in the same thread group on core 2 are marked as 1, the scheduler first checks whether there have threads in the *cur_uid* [1] thread group on core 2; if so, it chooses and schedules these threads, and if not, the leftmost record in the red-black tree will be scheduled. For the remaining two cores, it is similar to the above procedure. AGS guarantees that the threads corresponding to the same application are clustered into the same thread group, thus reducing the switching of the cache state and improving the cache hit rate to some extent. In this paper, we use AGS to perform centralized thread scheduling.

2.3.3.2   RS-Buddy System

Taking into account the energy efficiency, we use the *RS-buddy System* our group previously proposed to centrally allocate memory, so as to increase the idleness of the *Rank* and reduce the power consumption.

A memory *Rank* is a set of memory chips connected to the same chip selection, which is accessed simultaneously. In the *RS-buddy System*, the *Rank* identification information is added to the original buddy system to improve the idleness of the *Rank*. In the original buddy system, the consecutive free page blocks are organized in the *free_area* list and connected by a linked list. For example, *free_area* [2] corresponds to one or several 2^2 free pages, and *free_area* [10] corresponds to one or several 2^{10} free pages. If a thread requests size of 2^3 free pages, the algorithm firstly will call the *find_page* function to find it from *free_area* [3] and return if the pages are found. If not, the 2^4 free pages in *free_area* [4] will be removed and divided into two equal sizes. One part is allocated and the other part is inserted into *free_area* [3]. In the original buddy system, all the page blocks in the same *free_area*[] list are treated as the same level, and the physical memory is treated as a black box. To manage the memory states more efficiently, we modify the free list to a hierarchy: for the page block in the *free_area*[order] list, we reorganize the page block into a *Rank* [0..n] lists based on the *Rank* identification of the page block. During the

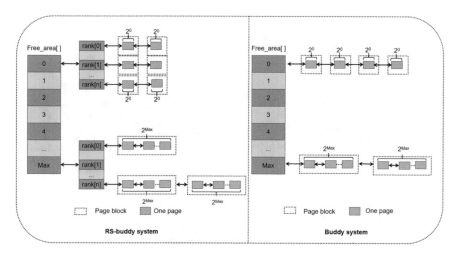

Fig. 2.3 Comparison between RS-Buddy and Buddy system

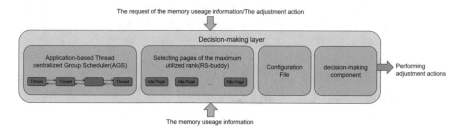

Fig. 2.4 Decision-Making layer

find_page process, the *RS-buddy System* will prefer selecting pages of the maximum utilized *Rank*. Figure 2.3 shows a comparison of RS-buddy and buddy system.

We introduce the RS-buddy page allocator to improve the memory usage efficiency, that is, to gather the memory usage into as few *Ranks* as possible, so as to better cooperate with the power mode management strategies. For example, the dynamic power state management policy mentioned in [8] controls the idle time of the chip by setting a certain threshold. When the idle time exceeds the threshold, the chip is dynamically adjusted to low power mode. If the memory usage is concentrated, it is possible to have as much of the memory as possible in the low power mode to achieve better power consumption reduction.

The function of the *Decision-Making layer* is shown in Fig. 2.4.

In the *Decision-Making layer*, if the information request sent from the *Interface layer* is received, the resource usage information is pulled from the information collection *layer* to this *layer*, and then the requested information is sent to the *Power Mode Adjustment Module* through the *Interface layer*. When the adjustment

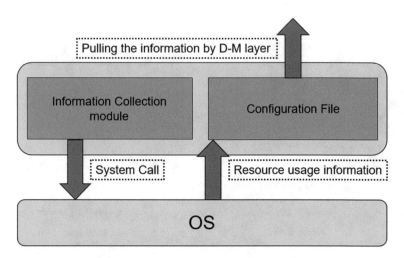

Fig. 2.5 The Implementation of Information-Collecting layer

action is sent from the *Interface layer*, the *Decision-Making* module implements the adjustment action according to the memory usage status after the AGS and RS-buddy work.

2.3.4 Information-Collecting Layer

We use the *Information-collecting layer* to get information about the resource usage. After the *Information-Collecting layer* receives the information-collecting request from the *Decision-Making layer*, we use hardware calls to get hardware resource usage and obtain the usage of CPU, cache, and memory regularly by looping calls. After that, the obtained information is stored in the configuration file. When the *Decision-Making layer* receives the information acquisition request, the required information is pulled from the module of the *Information-Collecting layer* (Fig. 2.5).

2.3.5 Interface Layer

This layer can be seen as an interface that we have reserved in combination with other advanced power mode adjustment strategies. The *Power Mode Adjustment Module* obtains the information required for power adjustment through the *Interface layer* in the framework, That is, the layer provides the service for information exchange. Figure 2.6 describes the functions assumed by the *Interface layer*.

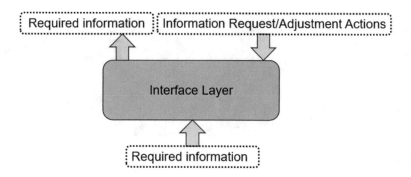

Fig. 2.6 The function of Interface layer

In the *Interface layer*, the resource usage information is transmitted. After the power management policy in the *Power Mode Adjustment Module* is determined, the resource usage information request is sent to the *Interface layer* from the *Power Mode Adjustment Module*, and then the *Interface layer* sends the request to the *Decision-Making layer*. The *Decision-Making module* in the *Decision-Making layer* returns the requested information to the *Power Mode Adjustment Module*. After the *Power Mode Adjustment Module* determines the power adjustment method, the adjustment action is sent to the *Interface layer*, and the *layer* sends the action to the *Decision-Making layer*.

We can easily combine different power adjustment strategies with the scheduler we use through the *Interface layer*. Different strategies can obtain the required information through the *Interface layer* to formulate their own adjustment strategy and then pass the adjustment method through the *Interface layer*. One implementation method we consider is to output the information that needs to be used to our specified file *F_Mem_Info*. The file is passed by the file location descriptor, and the file is finally read by the implementation of the *Interface layer*. And pass the information of the file to the *Power Mode Adjustment Module* through the data stream channel.

Here a pseudo code is used to show the implementation process of reading information at the *Interface layer*:

```
If information request
    If Mem_Info is not NULL
        Read the contents of the file.
        Feed the content to the Power Mode Adjustment Module.
ELSE waiting
End
```

The information required by the power adjustment strategy is passed through variables; we record the mapping between these variables and the variables used in the system to read resource usage, in the *Interface layer*. The information required

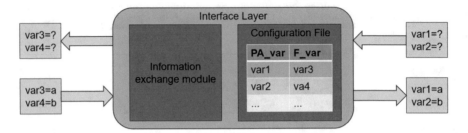

Fig. 2.7 Variable conversion

by the power adjustment strategy is passed through variables. We consider that there are some differences between the information acquisition methods of some power adjustment strategies and the methods in our framework. Therefore, when implementing information acquisition, it may be necessary to map between different variables. In the *Interface layer*, we record the mapping between the variables used in the power adjustment strategy and the variables used in our framework to read resource usage. After the *Power Mode Adjustment Module* issues a request for information, according to these mapping relationships, the variables in the request are mapped into corresponding variables used in the framework, and then the modified information request is sent to *Decision-Making layer*. Meanwhile, the *Interface layer* is also responsible for the information conversion. After reading the relevant file information, the *Interface layer* converts it into the variable name used by the previous power adjustment strategy and transmits it to the *Power Mode Adjustment Module*.

Figure 2.7 shows the process of variable query and conversion, where *PA_var* represents the variable used in the power adjustment strategy and *F_var* represents the variable used in the framework.

2.3.6 Workflow

The workflow of the framework is also shown in Fig. 2.1. To make better trade-off between system performance and power consumption, the *Power Mode Adjustment Module* periodically collects OS resource usage information, for instance, the utilization rate of the *Rank*, to formulate adjustment strategy to determined power management policy. Firstly, the information-collecting request is sent to the *Interface layer*. Then, it is passed by the *Decision-Making layer*. Finally, the *Decision-Making layer* pulls the required information from Information-Collecting *layer* and returns it to the *Power Mode Adjustment Module*. After obtaining the information, the *Power Mode Adjustment Module* formulates adjustment method based on the determined power mode management policy to adjust the power mode of the memory. It's worth noting that these adjustment actions are not performed in the *Power Mode Adjustment Module*. Instead, they are sent to the *Decision-Making layer* and executed in the *Decision-Making* component. This facilitates the choice between multiple strategies in the *Decision-Making layer* in follow-up study.

2.4   Experimental Verification

We use the *pandaboardES* development board which equipped with the OMAP4460 processor and installed Android 4.0.4 OS as our experiment platform. The OMAP4460 processor has two ARM Cortex A9s cores with a shared 1 MB L2 cache, and the highest frequency is up to 1GHz. The core instruction cache and data cache both are 32 KB. To appraise the performance improvement effect of our proposed strategy, we use the Android multi-threaded benchmark application [10].

The following figure depicts the implementation logic flow of our framework. For the core part of the framework (*Decision-Making layer*), firstly, the system determines if a request for information has been received. If not, it further determines whether the power adjustment strategy is completed, that is, if the power adjustment execution strategy is sent and performed by the *Decision-Making layer*. If the information acquisition request is received, the *Decision-Making layer* pulls the required information from the *Information Collection layer*. The AGS is used as the default scheduling policy in the *Decision-Making layer*. Figure 2.8 shows the processing logic of this layer.

For the *Power Mode Adjustment Module*, after obtaining the required data, the adjustment strategy is specified according to the dynamic power adjustment strategy. The difference from the strategies generated earlier is that after the policy is formulated, these adjustment actions are sent to the *Decision-Making layer*. The *Decision-Making layer* performs the adjustment action eventually. Figure 2.9 shows the processing logic of the power management module.

For the original AGS, the threads of the same application are centrally scheduled to reduce the cache state conversion to improve the response time, which may improve the system performance. However, in terms of energy efficiency, AGS is the same as most group schedulers. It only improves the efficiency of resource use, but there has been little progress in reducing power consumption and improving energy efficiency.

We combine the AGS with the *dynamic power state management* (DPM) algorithm mentioned in [8] and the BUT strategy mentioned in [16]; we named them AGS-DPM and AGS-BUT, respectively. At the same time, we use the *RS-buddy System* free page allocator to allocate the memory to increase the idleness of the *Rank*. More *Rank* idleness is achieved to reduce power consumption, which can effectively decrease unnecessary transitions between different power modes and cause reduced performance loss and energy cost.

Our experimental results are shown in Table 2.1. When the AGS policy is combined with the DPM strategy, compared with the original DPM, the performance is improved by 14.8%, the power consumption decreased by 9.9%, and the product

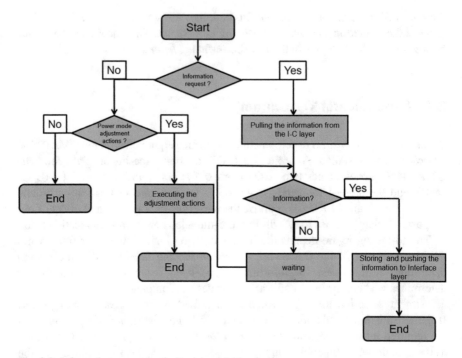

Fig. 2.8 Information processing in Decision-Making layer

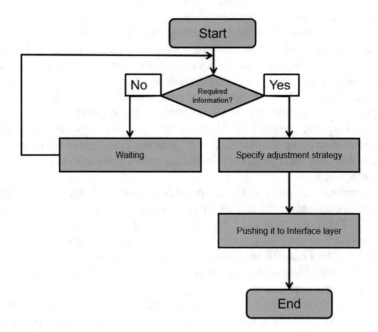

Fig. 2.9 The implement logic of Power Mode Adjustment Module

Table 2.1 Comparison with different power mode control policies

	Rank idleness	Transitions	Performance	Power	EDP
AGS-DPM	5.9%	−26.31%	14.8%	−9.9%	−20.1%
AGS-BUT	11.3%	−17.01%	9.5%	−13.2%	−22.5%

of energy delay (EDP) decreased by 20.1%. When it combined with the BUT strategy, the performance increased by 9.5%, the power consumption decreased by 13.2%, and the EDP decreased by 22.5%, compared with the original BUT strategy.

2.5 Conclusion

In this paper, we propose a framework to balance the performance and power consumption of embedded intelligence service devices. The framework adapts AGS [18] to achieve thread aggregation and combines it with the dynamic power mode control strategy together. The framework includes the *Information-Collecting layer*, the *Decision-Making layer*, the *Interface layer*, and the *Power Mode Adjustment Module*. The system collects statistics on the OS resource usage and transmits them through the information exchange component. This information is used by the power management strategy to formulate energy optimization method. To facilitate our subsequent research on selecting the optimal strategy in multiple strategies, the generated method is then executed by the *Decision-Making* component in the *Decision-Making layer*. In order to further improve energy efficiency, we also introduce a free pages allocator called *RS-buddy System* to centralize the allocation of memory. The *RS-buddy System* can gather the memory usage into as few *Ranks* as possible, thus improving the idleness of the *Rank*. The concentrated memory usage makes as much memory as possible in the low power mode, thus achieving better effect of reducing power consumption.

Experimental results show that when we combine AGS with DPM, the *Rank Idle Rate* is reduced by 5.9%, the power consumption is reduced by 9.9%, and the EDP is reduced by 20.1%. When the AGS and BUT strategies are combined, the *Rank Idle Rate* is reduced by 11.3%, the power consumption is reduced by 13.2%, and the EDP is reduced by 22.5%.

We are still working further to explore how to apply multiple power mode adjustment strategies to our framework so as to get better trade-off result.

Acknowledgments This work was supported by the National Science Youth Fund of Jiangsu Province (No. BK20190224).

References

1. A. El-Moursy, et al., Compatible phase co-scheduling on a CMP of multi-threaded processors. in *The 20th IEEE International Parallel & Distributed Processing Symposium,* Rhodes Island, Greece, 25–29 April 2006
2. A. Snavely, et al., Symbiotic job scheduling for a simultaneous multithreading processor. Paper presented at ASPLOS-IX Proceedings of the 9th International Conference on Architectural Support for Programming Languages and Operating Systems, Cambridge, MA, USA, November 12–15, 2000
3. D. Iorga, et al., A Portable Framework for Multi-core Interference Tuning and Analysis (2018). https://arxiv.org/pdf/1809.05197.pdf. Accessed 13 Sept 2018
4. F. Bellosa, M. Steckermeier, The performance implications of locality information usage in shared-memory multiprocessors. J Parallel Distrib. Comput. **37**(1), 113–121 (1996)
5. J. Philbin, et al., Thread scheduling for cache locality. in *ASPLOS-VII Proceedings – Seventh International Conference on Architectural Support for Programming Languages and Operating Systems*, Cambridge, Massachusetts, USA, October 1–5, 1996. (ACM Press, 1996), ISBN 0-89791-767-7
6. L. Barroso, et al., The datacenter as a computer. in *The Datacenter as a Computer: Designing Warehouse-Scale Machines, Third Edition Synthesis Lectures on Computer Architecture*, October 2018, p 189 (2009)
7. M. Shacklett, Edge computing: A cheat sheet (2017). https://www.techrepublic.com/article/edge-computing-the-smart-persons-guide/. Accessed 21 July 2017
8. R. Lebeck, et al., Power aware page allocation, Paper presented at ASPLOS-IX, 2000: Cambridge, MA, USA (2000)
9. R. L. McGregor, et al., Scheduling algorithms for effective thread pairing on hybrid multi-processors. Paper presented at the 19th IEEE International Parallel and Distributed Processing Symposium, Denver, CO, USA, 04–08 April 2005 (2005)
10. Roylongbottom, Android multithreading benchmark apps (2014). http://www.roylongbottom.org.uk/
11. S. Blagodurov, et al., Addressing Contention on Multicore Processors via Scheduling. in *Simon Fraser University, Technical Report 2009-16* (2009)
12. S. Zhuravlev, et al., Addressing shared resource contention in multicore processors via scheduling. Paper presented at ASPLOS-XV, 2010: Pittsburgh, Pennsylvania, USA (2010)
13. S. Yang, et al., EdgeCNN: Convolutional Neural Network Classification Model with small inputs for Edge Computing (2019). https://arxiv.org/pdf/1909.13522.pdf. Accessed 30 Sept 2019
14. S. Parekh, et al., Thread-sensitive scheduling for SMT processors.: Technical report, Dept. of Computer Science & Engineering, Univ. of Washington (2000)
15. D. Tam et al., Thread clustering: sharing-aware scheduling on SMP-CMP-SMT multiprocessors. ACM SIGOPS Oper. Syst. Rev. **41**(3), 47–58 (2007)
16. V. Delaluz, et al., Scheduler-based DRAM energy management, Paper presented at Design Automation Conference, New Orleans, LA, USA, USA, 10–14 June 2002 (2002)
17. Weissman, Performance counters and state sharing annotations: A unified approach to thread locality. Paper presented at ACM SIGOPS Operating Systems Review, 127–138 (1998)
18. Z. Zhu et al., Application-aware group scheduler for Android. Paper presented at the 5th International Conference on the Applications of Digital Information and Web Technologies: Bengaluru, India (2014)

Chapter 3
Formal Verification, Testing, and Inspection for Intelligent Services

Min Xu and Lisong Wang

3.1 Introduction

From self-driving cars to AlphaGo, artificial intelligence (AI) is progressing rapidly. Artificial intelligence makes our lives more convenient, but it also may bring us dangers. Just like Russia's president Vladimir Putin said: "Artificial intelligence is the future, not only for Russia, but for all humankind. It comes with enormous opportunities, but also threats that are difficult to predict. Whoever becomes the leader in this sphere will become the ruler of the world." So we should have a very convincing argument for its safety before applying an advanced intelligent system. How can we realize that argument is rigorously correct? Dijkstra said: "The only effective way to raise the confidence level of a program significantly is to give a convincing proof of its correctness" [8]. The answer is a mathematical proof. This is the reason why we need formal methods in AI. Formal methods are used to describe and analyze systems with a set of symbols and operations; depend on some mathematical methods and theories, such as algebra, logical, graph theory, or automata; and enhance the quality and safety of systems, so we call it formal. The properties of systems described formally can eliminate misunderstandings, and a system that satisfies its specification can be verified by formal techniques. Design or coding errors can be found in formal methods before deploying it to reduce the risk of damage of a system. This can help one to only care about the main properties of the system and can easily manage the complexity of the system. Formal methods include modeling, specification, verification and testing techniques. This chapter will present the four parts and finally give an example to illustrate.

M. Xu (✉) · L. Wang (✉)
Department of Computer Science and Technology, NUAA, Nanjing, China
e-mail: xumin@nuaa.edu.cn; wangls@nuaa.edu.cn

© Springer Nature Switzerland AG 2021
H. Gao, Y. Yin (eds.), *Intelligent Mobile Service Computing*, EAI/Springer
Innovations in Communication and Computing,
https://doi.org/10.1007/978-3-030-50184-6_3

33

3.2 Modeling

Modeling describes some properties of the system using mathematical methods. It keeps a little original details of the system by the process of abstraction. Here, a modeling formalism and a modeling language are distinct [3]. Formalisms belong to the field of mathematics composed of abstract syntaxes and formal semantics. Languages, such as NuXMV [22], are designed to implement formalisms. A language, generally, includes a comxiler, simulator, etc.

We choose a good formalism and language for formal verification, or model checking should depend on what type of system, what properties of system, etc. For example, we can formalize discrete time systems by using push-down automata [17] and finite state machines, concurrent processes by Petri nets [21] and communicating sequential processes (CSP) [16], and compositional modeling by reactive modules [2], process algebras [13], etc. The system model is verified by a specification of the properties of the system to be verified. Different specification languages deal with different properties. For example, specification languages, such as regular expressions, state charts diagrams, computation tree logic, etc., are used for reactive systems [19]. Here we will focus on behavior over time of reactive systems because these kinds of behavior often appear in intelligent systems. We will use a Kripke structure [18] to define reactive systems' behavior.

3.2.1 Kripke Structure Modeling

A Kripke structure consists of a nonempty finite set W named state, a set of relations T is a subset of $W \times W$, and a function F, if each state with a set of propositions of this state are true, then the value of F is 1 else is 0. A path in this model M from a state $w \in W$ is an infinite sequence of states $p = w_0 w_1 w_2 \ldots w_k w_{k+1} \ldots$. such that $w_0 = w$ and $R(w_i, w_{i+1})$ holds for all $k \geq 0$. Here, let $Atom\,P$ be a set of atomic propositions. So a Kripke structure M over $Atom\,P$ can be represented as a three-tuple $M(W, T, F)$.

In this chapter, we only cover the first-order logic. So here we just use the logical connectives such as not $\neg$, and $\wedge$, or $\vee$, implies $\rightarrow$, and quantifiers such as a universal quantifier($\forall$) and an existential quantifier ($\exists$).

Let $V = \{v_1, \ldots, v_n\}$ be a variable set in a system. The range of a variable in V is a finite set D. A valuation for V is a mapxing (or function) from V to D. We can assign each variable in V a value in D to represent a state $w \in W$ of a concurrent system. So a valuation $w : V \rightarrow D$ is a state w. Generally, we can use a logical formula to represent a valuation w; in other words, the valuation makes the formula true. For example, given a variable set $\{v_1, v_2, v_3, v_4\}$ and a valuation$\{w(v_1) = 10, w(v_2) = 7, w(v_3) = 9, w(v_4) = 1\}$, the corresponding formula is $(v_1 = 10) \wedge (v_2 = 7) \wedge (v_3 = 9) \wedge (v_4 = 1)$, so, here, we use w to denote the formula. Now we know we can use a formula to represent a state in a system.

For a transition in a system, we can use an ordered pair $< w, w' >$ to denote, where w' is next state in path p, then we can use $T(w, w')$ to represent.

Now we can use the first-order formula to explain a Kripke structure $M = (W, T, F)$ that represents the concurrent system.

1. W is a set of logic formulas.
2. Letting w and w' be two states, then $T(w, w') \in T$ if w is $True$ when each $v \in V$ is assigned the value $w(v)$ and the same to w'.
3. The function $F : W \rightarrow 2^{AtomP}$ is defined so that $F(w)$ which is the subset of all atomic propositions is $True$ in w. If the range of v is $\{True, False\}$, then $v \in F(w)$ illustrates $w(v) = True$, and $v \notin L(s)$ means $w(v) = False$.

3.2.2 An Example

To show how to use the notions in this section, we give a simple example.

This is a program to implement a summing procedure from 1 to 5, which means we want to get the value of $1 + 2 + 3 + 4 + 5$. sum variable denotes the sum of the value, variable i denotes the value from 1 to 5, and pc denotes the program counter.

$p_1 : sum := 0;$
$p_2 : i := 1;$
$p_3 : while\ i <= 5\ do$
$p_4 : sum := sum + i;$
$p_5 : i := i + 1$
$p_6 : end$

So for the variables $\{sum, i, pc\}$, the range of sum is an integer, denoted by Z; the range of i is the naturals, denoted by N; and the range of pc is $\{p_1, p_2, p_3, p_4, p_5, p_6\}$, denoted by PC.

The Kripke structure $M = (W, T, F)$ in the program can be represented by the following:

1. $W = Z \times N \times PC$.
2. $T = \{((sum = 0) \wedge (PC = p_1), (sum = 0) \wedge (i = 1) \wedge (PC = p_2)), ((sum = 0) \wedge (i = 1) \wedge (PC = p_2)), (sum = 0) \wedge (i = 1) \wedge (PC = p_3)), ((sum = 0) \wedge (i = 1) \wedge (PC = p_3)), (sum = 1) \wedge (i = 1) \wedge (PC = p_4))\ldots\}$.
3. $L((sum = 0) \wedge (PC = p_1), (sum = 0)) = \{sum = 0, PC = p_1\}, L(((sum = 0) \wedge (i = 1) \wedge (PC = p_2))) = \{sum = 0, i = 1, PC = p_2\}, L(((sum = 0) \wedge (i = 1) \wedge (PC = p_3))) = \{sum = 0, i = 1, PC = p_3\}, \ldots$.

The path in the this program that starts in an initial state is $((sum = 0) \wedge (PC = p1))((sum = 0) \wedge (i = 1) \wedge (PC = p2))((sum = 0) \wedge (i = 1) \wedge (PC = p3))((sum = 1) \wedge (i = 1) \wedge (PC = p4))((sum = 1) \wedge (i = 2) \wedge (PC = p5))((sum = 1) \wedge (i = 2) \wedge (PC = p3))((sum = 3) \wedge (i = 2) \wedge (PC = p4)) \cdots$.

3.3   Specification

Specification is usually described in some mathematical or logical methods. Generally, it uses temporal logic for hardware and software systems, which can assert what
behaviors of the system over time. Using these methods to specify the design of a
project can keep the consistency between the different modules of a project during
its development and maintenance. Only if we give the specification of a program,
we can say a program is correct or not. When a system has specification, verification
of a system is necessary. A contract between the developer and the customer can be
described by formal specifications.

Now we introduce briefly the classical temporal logic to describe formal
specification. In a reactive system, temporal logics describe formally sequences of
transitions between states. In the temporal logics, time is not mentioned explicitly.
Instead, we often say that eventually or possibly some designated states have
reached a formula or not and do not mention time explicitly. So eventually or
possibly, they are specified using special temporal operators. These operators can
be combined with logical connectives. Temporal logics provide the semantics of
those operators. Here, we will use a powerful temporal logics called CTL* [7, 11].

3.3.1   The Computation Tree Logic (CTL*)

The possibility of transition of a system can form a tree structure called computation
tree. Properties of computation trees can be described by CTL* formulas. The root
of the tree is the initial state in Kripke structure of the system and extends to be an
infinite tree. The possible executions starting from the initial state can be showed by
the computation tree.

In CTL* formulas are composed of propositional variables, logical constant,
connectives, temporal operators, and path quantifiers. The branching structure in
the computation tree is described by the path quantifiers. Path quantifiers are **A**
and **E**. **A** denotes for all computation paths which specifies all of the paths have
some property. **E** denotes for some computation path which specifies some of the
paths have some property. There are five basic temporal operators which describe
properties of a path through the tree:

1. $\Box$ means "always" which specifies that each state on the path has the property.
2. $\Diamond$ means "eventually" which some state on the path in the future will have the
   property.
3. $\bigcirc$ means "next time" which specifies the next state on the path has the property.
4. U means "until" which specifies the states all hold the first property until some
   state holds the second property.
5. $\lor$ means "release" which specifies when the first property is released by the
   states, the second property is held by all the states left.

CTL* has two types of formulas. One is state formulas which are true in a specific state. Another is path formulas which are true along a specific path. $Atom\,P$ is the set of atomic proposition. The following rules form the syntax of state formulas:

1. p is a state formula if $p \in Atom\,P$.
2. $\neg p$, $p \vee q$, and $p \wedge q$ are state formulas if p and q are also state formulas.
3. $\mathbf{A}\,p$ and $\mathbf{E}\,p$ are state formulas if p is a path formula.

Two additional rules are needed to specify the syntax of path formulas:

1. p is a path formula if p is a state formula.
2. $\neg p$, $p \vee q$, $p \wedge q$, $\Box p$, $\Diamond p$, $\bigcirc p$, $p \cup q$, and $p \vee q$ are path formulas if p and q are also path formulas.

The set of state formulas of CTL* is generated by the above rules.

Now we use a Kripke structure to define the semantics of CTL*. A Kripke structure M is (W, T, F), where W is the set of states; T is the relation which is a subset of $W \times W$, which satisfies refection, which means $\forall w \in W \Rightarrow (w, w) \in T$; and $F : W \to 2^{Atom\,P}$ is a function that illustrates a set of atomic propositions is true in that state. A path p in M is an infinite sequence of states, $p = w_0 w_1 \cdots w_i w_{i+1} \cdots$, $\forall i\ i \geq 0, (w_i, w_{i+1}) \in T$.

The suffix of ξ starting at w_i is denoted by ξ^i. If p, s are state formulas, the notation $M, s \vDash p$ means state s can imply state p in the Kripke structure M. Similarly, $M, \xi \vDash f$ means that along path ξ can imply the path formula f in the Kripke structure M if f is a path formula. We define inductively the relation $\vDash$ as follows: (here, p is an atomic proposition, p_1 and p_2 are state formulas, and f_1 and f_2 are path formulas):

1. $M, s \vDash p$ iff $p \in F(s)$.
2. $M, s \vDash \neg p$ iff $M, s \nvDash p$.
3. $M, s \vDash p_1 \vee p_2$ iff $M, s \vDash p_1\ or\ M, s \vDash p_2$.
4. $M, s \vDash p_1 \wedge p_2$ iff $M, s \vDash p_1\ and\ M, s \vDash p_2$.
5. $M, s \vDash \mathbf{E} f_1$ iff there exists a path ξ from $s \Rightarrow M, s \vDash f_1$.
6. $M, s \vDash \mathbf{A} f_1$ iff for all path ξ starting from $s \Rightarrow M, s \vDash f_1$.
7. $M, \xi \vDash p_1$ iff s is the first state of $\xi \Rightarrow M, s \vDash f_1$.
8. $M, \xi \vDash \neg f_1$ iff $M, \xi \nvDash f_1$.
9. $M, \xi \vDash f_1 \vee f_2$ iff $M, \xi \vDash f_1\ or\ M, \xi \vDash f_2$.
10. $M, \xi \vDash f_1 \wedge f_2$ iff $M, \xi \vDash f_1\ and\ M, \xi \vDash f_2$.
11. $M, \xi \vDash \bigcirc f_1$ iff $M, \xi^1 \vDash f_1$.
12. $M, \xi \vDash \Diamond f_1$ iff there exists a $k \geq 0 \Rightarrow M, \xi^k \vDash f_1$.
13. $M, \xi \vDash \Box f_1$ iff $\forall i \geq 0, M, \xi^i \vDash f_1$.
14. $M, \xi \vDash f_1 \cup f_2$ iff $\exists k\ k \geq 0,\ M, \xi^k \vDash f_2\ and\ \forall ji \leq j < k, M, \xi^j \vDash f_1$.
15. $M, \xi \vDash f_1 \vee f_2$ iff $\forall j,\ j \geq 0, \forall i,\ i < j, M, \xi^i \nvDash f_1 \Rightarrow M, \xi^j \vDash f_2$.

Note: any other CTL* formula can be expressed by the operators $\vee, \neg, \bigcirc, \cup,\ and\ \Diamond$.

3.3.2 Fairness

In many cases, we want to keep not only correctness but also fair computation paths. For example, we may want to consider some server protocols which can provide reliable server which have the property that no client ever continuously submits requests but never respondence. The semantics of the logic is named the fair semantics. A set of states can describe a fairness constraint by a formula of the logic. A fair path should include a state of each fairness constraint infinitely when we use sets of states to represent fairness constraints.

A fair Kripke structure can be described by four-tuple $M = (W, T, F, V)$, where W, T, and F are the same as we defined before and $V \subseteq 2^{Atom P}$ is a set which describe fairness constraints. Let $\xi = w_0 w_1, \cdots, w_i w_{i+1}, \cdots$ be a path in M. It can be defined by $inf(\xi) = \{w | w = w_i \text{ for infinitely many } i\}$.

If the path ξ is a fair path, it should satisfy the following condition:

$$\xi \text{ is a fair } \Leftrightarrow \forall P \; P \in F, \; inf(\xi) \cap P \neq \emptyset.$$

The semantics of an ordinary Kripke structure of CTL* can describe the semantics of an ordinary Kripke structure of CTL*. In states of the fair Kripke structure M, we use $M, s \vDash_V p$ to represent that the state formula p is true and $M, s \vDash_V f$ to represent that the path formula f is true along path ξ. The semantics of an ordinary Kripke structure can change just only clauses 1, 5, and 6 in the original semantics.

1. $M, s \vDash_V p$ iff $\exists$ a fair path ξ starting from w and $p \in F(s)$.
5. $M, s \vDash_V \Diamond f_1$ iff $\exists$ a fair path ξ from s s.t. $M, s \vDash_V f_1$.
6. $M, s \vDash_V \Box f_1$ iff $\forall$ fair path ξ starting from s s.t. $M, s \vDash g_1$.

We can use server protocols for reliable server to illustrate the use of fairness. Here we use only one fairness constraint for each client that illustrates the reliability of that server. The fairness constraint is $\forall client_i (\neg request(client_i) \lor respondence(client_i))$. So, a computation path ξ is fair $\Leftrightarrow \forall client_i (\neg request(client_i) \lor respondence(client_i)) \cap inf(\xi) \neq \emptyset$.

The more details can be found in [14].

3.4 Verification

In the development phase, we can evaluate our work whether it satisfies the specification of the requirements of that phase or not through verification. In other words, verification can help us to determine if the work we have done meets the requirements and specifications. At present, verification consists of two methods: one is deductive verification and the other is model checking.

3.4.1   Deductive Verification

Using axioms and rules to prove whether systems are correct or not is called deductive verification. Most of computer scientists believe deductive verification is very important. So we have used it in many fields of software development. Deductive verification can be automated to prove infinite-state systems. However, deductive verification needs much knowledge of mathematics; thus, only experts can utilize it. Moreover, deductive verification is time-consuming to prove some little scale problem. Finally, when we cannot prove the system is correct, we cannot also prove that the system is wrong. So here we will only emphasize model checking.

3.4.2   Model Checking

We can use a technique of model checking for automation to verify finite state systems. General approaches search exhaustively the state of system to check if specification of the system has been met or not. This procedure always will stop and give us an answer that the system is true of false if we have sufficient memory and time.

We can easily describe the model checking problem with Kripke structure and a temporal logic formula. Using temporal logic formula p to represent some specification and Kripke structure $M = (W, T, F)$ to describe the system, then we can check $M, s \vDash p$ or not. There are initial states in the system, and model checking will check whether the system satisfies these specifications according to the initial states.

3.4.3   Symbolic Model Checking

In 1987, McMillan [4, 20] verified much larger systems through the state transition graphs of symbolic representation. They utilized Bryant's ordered binary decision diagrams (OBDD) [5] to construct the new symbolic representation. OBDDs is a very efficient algorithm which provides a canonical form for Boolean formulas. Symbolic representation can describe the rules in the state space according to the specifications, so it can verify systems whose scale is larger than algorithms described by handled explicit state. By the new representation, the original CTL model checking algorithm [7] can verify some systems whose states are more than 10^{20} states.

3.4.3.1 Fixed-Point Representations

A Kripke structure $M(W, T, F)$ is finite, where W is a finite set and the power of W is denoted by $P(W)$. Let $\pi : P(W) \to P(W)$ be a transform or a function.

1. π is monotonic if $R \subseteq S$ implies $\pi(R) \subseteq \pi(S)$;
2. π is $\cup$ continuous if $R_1 \subseteq R_2 \subseteq \cdots \subseteq R_i \subseteq R_{i+1} \subseteq \cdots \Rightarrow \pi(\cup_i(R_i)) = \cup_i(R_i)$;
3. π is $\cap$ continuous if $P_1 \supseteq P_2 \supseteq \cdots \supseteq P_i \supseteq P_{i+1} \supseteq \cdots \Rightarrow \pi(\cap_i P_i) = \cap_i \pi(P_i)$;

$$\pi^i(W) = \underbrace{\pi(\cdots(\pi(\pi(W))))}_{i}.$$

The recursive definition is the following:

$$\begin{cases} \pi^0(W) = W \qquad n = 0 \\[2mm] \pi^{i+1}(W) = \pi(\pi^i(W)) \; n = i \end{cases}$$

There are least fixed point denoted by $least(\pi(W))$ and greatest fixed point denoted by $great(\pi(W))$ on a monotonic function, just like the following:

$$\begin{cases} least(\pi(W)) = \cap\{W \mid \pi(W) \supseteq W\} \\[2mm] great(\pi(W)) = \cup\{W \mid \pi(W) \subseteq W\} \end{cases}$$

When π is monotonic and *cup* or *cap* is continuous, the definition is as follows:

$$\begin{cases} least(\pi(W)) = \cap_i \pi^i(True) \; \cap \, continuous \\[2mm] great(\pi(W)) = \cup_i \pi^i(False) \cup continuous \end{cases}$$

We can get some lemmas defined on finite Kripke structures [7, 12].

Lemma 1 π *is* $\cup$ *continuous and* $\cap$ *is continuous if* W *is finite and* π *is monotonic.*

Lemma 2 $\forall i \; \pi^i(False) \subseteq \pi^{i+1}(False)$ *and* $\pi^i(True) \supseteq \pi^{i+1}(True)$ *if* π *is monotonic.*

Lemma 3 $\exists n_0 \in Z, s.t. \; least(\pi(W)) = \pi^{n_0}(False).$ *and* $\exists m_0 \in Z, s.t. \; great(\pi(W)) = \pi^{m_0}(Ture)$ *If* W *is finite and* π *is monotonic, where* Z *is an integer set.*

We can use a least or greatest fixed point to define CTL operators when we use $\{w | M, w \vDash p\}$ in $P(W)$ to present CTL formula g [10].
g_1, g_2 are CTL formulas.

- $A\Diamond g_1 = least(g_1) \vee A \bigcirc W$
- $E\Diamond g_1 = least(g_1) \vee E \bigcirc W$

- $\mathbf{A}\Box g_1 = great(g_1) \wedge \mathbf{A} \bigcirc W$
- $\mathbf{E}\Box g_1 = great(g_1) \wedge \mathbf{E} \bigcirc W$
- $\mathbf{A}[g_1 \cup g_2] = least(g_2) \vee (g_1 \wedge \mathbf{A} \bigcirc W)$
- $\mathbf{E}[g_1 \cup g_2] = least(g_2 \vee (g_1 \wedge \mathbf{E} \bigcirc W)$
- $\mathbf{A}[g_1 \vee g_2] = great(g_2) \wedge (g_1 \vee \mathbf{A} \bigcirc W)$
- $\mathbf{E}[g_1 \vee g_2] = great(g_2) \wedge (g_1 \vee \mathbf{E} \bigcirc W)$

We can easy to know least fixed points correspond to final event and greatest fix points correspond to hold properties forever. Thus, $\mathbf{A}\Diamond g_1$ has a least fixed point and $\mathbf{A}\Box g_1$ has a greatest fixed point.

3.4.3.2 Symbolic Model Checking for CTL

We use OBDD to represent the Kripke structures and the logic of quantified Boolean formulas (QBF) [1, 15] to denote operations on Boolean formulas.

The set of formulas $QBF(V)$ can be defined as follows:
$V = \{v_0, \cdots, v_{n-1}\}$ is a set to represent propositional variables:

- $v \in V$ is a formula;
- $\neg f_1, f_1 \wedge f_2, f_1 \vee f_2$ are formulas if f_1 and f_2 are also formulas,
- $\exists v f$ and $\forall v f$ are formulas, if f is a formula and $v \in V$.

We use a function $\pi : V \to \{0, 1\}$ to represent a truth assignment. Here we introduce the notation $\pi_{(a \to v)}$ to represent the truth assignment when $a \in \{0, 1\} = V$. So it can be defined by the following:

$$\pi_{(a \to v)}(v') = \begin{cases} a & \text{if } v = v' \\ \pi(v') & \text{otherwise.} \end{cases}$$

$\pi \vDash f$ denotes that f is true if $\pi(f) = 1$, where f is a formula and π is a truth assignment. The definition of $\vDash$ is the following:
f, f_1, f_2 are formulas in $QBF(V)$.

- $\pi \vDash v \Leftrightarrow \pi(v) = 1$
- $\pi \vDash \neg f \Leftrightarrow \pi \nvDash f$
- $\pi \vDash f_1 \vee f_2 \Leftrightarrow \pi \vDash f_1 \vee \pi \vDash f_2$
- $\pi \vDash f_1 \wedge f_2 \Leftrightarrow \pi \vDash f_1 \wedge \pi \vDash f_2$
- $\pi \vDash \exists v f \Leftrightarrow \pi_{(a \to 0)} \vDash f \vee \pi_{(a \to 1)} \vDash f$, and
- $\pi \vDash \forall v f \Leftrightarrow \pi_{(a \to 0)} \vDash f \wedge \pi_{(a \to 1)} \vDash f$.

Relational product operations exist in which quantifiers can be represented by $\exists x[f_1(x, y) \wedge f_2(x, y)]$ generally.

3.4.3.3 Algorithm of Model Checking

The algorithm of model checking can be implemented by a procedure or function name *module_checker*. The input of *module_checker* is the CTL formula to be checked, and the output of *module_checker* is an OBDD [9]. The definition is the following: a is a proposition f, f_1, f_2 are formulas.

- *module_checker*$(a) = \{v \in V | \pi(v) = a\}$,
- *module_checker*$(f_1 \wedge f_2) = $ *module_checker*(*module_checker*$(f_1) \wedge$ *module_checker*(f_2))),
- *module_checker*$(\neg f) = \neg$*module_checker*(*module_checker*(f),
- *module_checker*$(\mathbf{E} \bigcirc f) = $ *module_checker*$\mathbf{E} \bigcirc$ (*module_checker*(f)),
- *module_checker*$(\mathbf{E}[f_1 \cup f_2]) = $ *module_checker*$\mathbf{E}\cup$
  (*module_checker*(f_1), *module_checker*(f_2)),
- *module_checker*$(\mathbf{E}\square f) = $ *module_checker*$\mathbf{E}\square$(*module_checker*(f)).

module_checker$\mathbf{E}\bigcirc$ means if the state has a successor in f which is true, the formula $\mathbf{E} \bigcirc f$ is true. In other words,

$$module\_checker \mathbf{E} \bigcirc (f(v)) = \exists v'[f(v') \wedge T(v, v')].$$

Where $T(v, v')$ is a relation in OBDD.
module_checker$\mathbf{E}\cup$ can use the least fixed point to compute

$$\mathbf{E}[f_1 \cup f_2] = least(f_2) \vee (f_1 \wedge (\mathbf{E} \bigcirc W)).$$

The other formula can be processed similarly.

3.4.4 Fairness in Model Inspecting

For fairness, we assume to use CTL formulas $Con = \{c_1, c_2, \cdots, c_n\}$ to represent fairness constraints. We can define procedure *module_checker Fair* to check whether formulas f of specifications satisfy the Con or not.

Fairness constraints $\mathbf{E}\square c$ means there is a path which holds all c from the start to infinity. And the state W makes c true which has the following properties:

1. $\forall w \in W \Rightarrow f = True$,
2. $\exists p(f = True \; \forall w \in p)$

where p is a path in CTL.
By means of a fixed point, we can represent symbolic model checking as follows:

$$\mathbf{E}\square f = great(f) \wedge \bigwedge_{k=1}^{n} \mathbf{E} \bigcirc \mathbf{E}[f \cup (W \wedge P_k)]$$

So under fairness constraint, $module\_checker\,Fair\,E\square$ can be computed as follows:

$$module\_checker\,Fair\,E\square(f) = great(f) \wedge \bigvee_{i=1}^{n} \mathbf{E} \bigcirc (\mathbf{E} \cup (f, W \wedge p_i))$$

$module\_checker\,Fair\,E\bigcirc$, $module\_checker\,Fair\,E\cup$ can be computed similarly.

3.5   Testing

Software testing describes a process used to facilitate the qualification, integrity, security, and quality of software. In other words, software testing is a review or comparison process between actual output and expected output according to the specification. Testing methods can be applied on an actual system directly rather than a model and can deal with infinite-state systems. But testing cannot cover all the possible execution cases of a system, just depends on some criteria.

3.5.1   Software Testing Method

The testing methods mainly include white box testing, black box testing, and gray box testing from the perspective of whether you care about the internal structure or specific implementation of the software. White box testing methods mainly include code inspection method, static quality measurement method, logical coverage method, basic path test method, domain test, symbol test, path coverage, etc. Black box testing methods mainly include equivalence class division method, boundary value analysis method, error inference method, causality diagram method, decision table drive method, orthogonal experiment design method, function diagram method, etc.

The test methods can be divided into static tests and dynamic tests from the perspective of whether to execute the program. Static tests include code inspection, static structural analysis, code quality metrics, etc. Dynamic testing consists of three parts: constructing test cases, executing programs, and analyzing the output of programs.

Testing has different stages as follows:

- Unit (module) testing. The unit test is mainly to test the module of the software and find out that the actual function of the module does not conform to the specification and coding errors. Because the module is small in scale, single in function, and simple in structure, the test methods adopted are static test method and white box test.

- Integration testing. Integration testing is the second phase of software testing. At this stage, modules that have been assembled in strict accordance with program design requirements and standards are usually tested simultaneously to clarify the correctness of the assembly of the program structure and to discover problems related to the interface. At this stage, a combination of white box and black box is generally used for testing to verify the rationality of the design at this stage and the realization of required functions.
- System testing. System tests check whether the system meets the software requirements. The main test content in this phase includes robustness test, performance test, function test, installation or anti-installation test, user interface test, stress test, reliability and safety test, etc. The system test mainly uses the black box method for testing.
- Validation testing. Validation testing is the testing work to be performed before the software product is put into actual execution. Compared with system test, validation testing differs from testers only, and validation test is performed by the user. The main goal of validation testing is to show users that the software developed meets predetermined requirements and relevant standards and to verify the effectiveness and reliability of the software's actual work and to ensure that users can successfully use the software to complete established tasks and functions.

Now we just briefly introduce what are white box testing and black box testing as end of the section.

Black box testing, as its name implies, simulates a software testing environment as an invisible "black box." Observe the data output through data input and check whether the internal function of the software is normal. When the test is unfolded, data is entered into the software and waits for data to be outputted. If the data output is consistent with the expected data, it proves that the software passes the test. If the data is different from the expected data, even if the difference is small, it also proves that there is a problem in the software program, and it needs to be resolved as soon as possible.

Compared with black box testing, white box testing has a certain degree of transparency. The principle is to debug the internal working process of the product according to the software's internal applications and source code. During the testing process, it is often analyzed in collaboration with the internal structure of the software. The biggest advantage is that it can effectively solve the problems of internal applications of the software. It is often combined with the black box test method during the test. The test method can also effectively debug such situations. Among them, the judgment test is one of the most important test program structures in the white box test method. Such a program structure, as an overall implementation of the program logic structure, has a more important role for the program test. This type of testing method covers all types of code in the program, and covers a wide range, which is suitable for multi-type programs. In actual detection, the white box test method is often used in combination with the black box test method. Take the unknown error detected in the dynamic detection method as an example. First, use the black box test method. If the program input data is the same as the output data, the internal data is not. If there is a problem, it should be analyzed from the code side. If there is a problem, use the white box test method to analyze the internal structure of the software until the problem is detected and amended in time.

3.6 Model Checking in Practice

3.6.1 The NuXMV Model Checker

NuSMV [6] is a classic model detection tool, which implements symbol model detection technology efficiently. NuSMV uses BDD to alleviate and achieve the state explosion problem, has a good software architecture, and is easy to customize and extend. NuXMV inherits all the functionalities of NuSMV and extends to specify the infinite-state systems.

In this part, we will introduce the NuXMV grammar and how to use it to check an intelligent system which should satisfy some properties.

3.6.2 Grammar of NuXMV

NuSMV uses its language to describe Kripke's structure and special verification specifications. The Kripke structure is often called finite-state machine (FSM) in NuXMV [22]. NuXMV has two useful expressions: *init* expression and *next* expression. The *init* expression is used to describe the initial state, and the *next* expression is used to describe the transition relationship. Programs written in NuXMV are often called smv programs. The smv program consists of modules. The types of state variables are very similar to other computer language, such as C, Java, Python, etc. Here do not provide to describe detailed, these can be found in the paper [22].

3.6.2.1 MODULE

A module consists of a module name and a module definition, and a module definition consists of a parameter and a body. The main part of the module is divided into three categories: Variables, Constraint, and Specification. The Variables section is used to describe the state set of the Kripke model; the Constraint section is used to describe the transition relationship of the Kripke model and some restrictions on the model; and the Specification section is used to describe the specification of special verification. The smv program must have at least one module called main, and the main module cannot have formal parameters. Multiple module descriptions can be used to describe the FSM and then combined into a whole FSM.

The following is an example 1:

```
1    MODULE main
2    VAR
3     s : boolean;
4    ASSIGN
5     init(s)  := FALSE;
6     next(s)  := TRUE;
7    CTLSPEC
8          EX s = TRUE
```

where "VAR i: boolean" denotes i is a variable and its type is boolean, "ASSIGN" belongs to the constraint which describes how a system works, 'init(i) := FALSE' denotes i initial value is FALSE, "next(i) := TRUE" denotes its value in the next state is TRUE, "CTLSPEC" introduces a formula in CTL to describe specification of the system, and **EX** means the system exists in the next state whose value is TRUE.

3.6.2.2 Types and Variables

NuXMV provides many data types, such as Boolean, integer, enumeration, word, and arrays types. Variables in NuXMV are declared by **VAR** which describe the states of a system. The definition of the variable of form is $< state\_variable\_name >:< data\_type >;$, where $< state\_variable\_name >$ denotes the name of the variable and $< data\_type >;$ denotes the type of data; in general, we will add a semicolon at the end of the statement. The second and third lines in example 1 show how to apply (VAR) to define variables.

Some data type are introduced by the following:

- Boolean Type: symbolic values **FALSE** and **TRUE**,
- Enumeration Types: full enumerations of all the values, for example, {SUCCESS, 1, 5, PASS},
- Integer: positive or negative integer number,
- Real: the rational numbers,
- Array: for example, array 0.5 of integer: 0 is lower bound, 3 is upper bound for the index, and integer is the type of the elements in the array,
- ...

NuXMV assigns values to variables by the keyword **ASSIGN**. We can use **init** to assign the initial values of the state in the system and **next** to assign the value of the next state of the system. Lines 4, 5, and 6 in example 1 show how to apply **ASSIGN** to assign values to variables.

There are two important expressions, Case Expression and If-Then-Else Expression, in NuXMV.

The syntax of If-Then-Else Expression is $<$ *bool_expr* $>$? $<$ *expr*1 $>$:$<$ *expr*2 $>$, where $<$ *bool_expr* $>$ must be a Boolean expression and $<$ *expr*1 $>$ and $<$ *expr*2 $>$ are any expressions; if $<$ *bool_expr* $>$ is true, then we can get the value of $<$ *expr*1 $>$; else get the value of $<$ *expr*2 $>$.

The syntax of Case Expression is the following:

```
case
condition1 : expression1;
condition2 : expression2;
. . .
1 :          expression N
esac
```

When the first value of condition k is true, the Case Expression returns the value of the kth expressionK on the right-hand side of "$:$ $,$" where "1" means other cases.

3.6.2.3   Specifications

NuXMV provides linear temporal logic (LTL), computation tree logic (CTL), and property specification language (PSL) to check whether the system satisfies the specification or not. NuXMV use **CTLSPEC**, **LTLSPEC**, and **PSLSPEC** to insert formulas of specifications to check.

NuXMV represent differently five basic temporal operators in CTL in Sect. 3.3.1. Refer to the paper [22] for details.

1. **G** denotes $\square$,
2. **F** denotes $\diamond$,
3. **X** denotes $\bigcirc$,
4. **U** denotes $\cup$.

Lines 7 and 8 in example 1 show how to apply **CTLSPEC** to check.

3.6.2.4   Module and Program

There must be a module named main in the program of NuXMV, which is just the main function in C language program. The other modules, in the program of NuXMV, are similar to general functions. Now we can run example 1. Because the example satisfy the specification **EX**s = **TRUE**, means the system should exist a next state is TRUE, obviously, it is correct. So we can obtain the result as following:

```
-- specification EX s = TRUE   is true
```

If we change the specification to the **EG** s = **TRUE** that means there exist a path holds s always true in the all states of the path, here, just a state s and the initial value of s is **FALSE**. So it is impossible in the example; the NuXMV gives the counterexample as follows:

```
Trace Type: Counterexample
  -> State: 1.1 <-
     s = FALSE
```

3.6.3 An Example

Now general intelligent services are very complex and concurrent and to know that the system has properties that we need, formal methods are a good way to help us. Here an example has been illustrated.

Now we consider that there are two Agents that can entertain in one place, but this place can only be accessed by one Agent at any time. Now we design a program that allows both Agent1 and Agent2 to have the opportunity to enjoy this place. Let each Agent to have four states: sleeping, trying, enjoying, and exiting. The sleeping state indicates that the Agent is idle, the trying state indicates that the Agent wants to enter this place to entertain, the enjoying state indicates that the Agent is entertaining, and the exiting state indicates that the Agent wants to leave this place. If only one Agent wants to go to this place while trying, then it can go. When both Agents are in the trying state and both want to go, then set a variable *semaphore* type to Boolean. If the *semaphore* value is **FALSE**, Agent1 can go in and change the value of semaphore to **TRUE**. If the semaphore value is **TRUE**, Agent2 can go in and change the value of semaphore to **FALSE**. To describe the syntax of NuXMV in more detail, consider the following program:

```
1   MODULE main
2   VAR
3     state_agent1: {sleeping, trying, enjoying,exiting};
4     state_agent2: {sleeping, trying, enjoying,exiting};
5     semaphore : boolean;
6     ag1: process agent(state_agent1, state_agent2, semaphore, FALSE);
7     ag2: process agent(state_agent2, state_agent1, semaphore, TRUE);
8
9   ASSIGN
10    init(semaphore) :=  FALSE;
11
12  CTLSPEC
13    EF(( state_agent1 = enjoying) & (state_agent2 = enjoying))
14  CTLSPEC
15    AG(( state_agent1 = trying) -> AF(state_agent1 = enjoying))
16  CTLSPEC
17    EX( state_agent1 = trying) & EX(state_agent2 = trying)
18
19  MODULE  agent(state0, state1, semaphore, semaphore0)
20  ASSIGN
21     init(state0) := sleeping;
22     next(state0) :=
23         case
24           (state0 = sleeping) : {trying, sleeping};
25           (state0 = trying) & (state1 = sleeping) : enjoying;
26           (state0 = trying) & (state1 = trying) &
```

```
27              (semaphore = semaphore0) : enjoying;
28              (state0 = enjoying) : {enjoying, exiting};
29              (state0 = exiting) : {sleeping};
30              TRUE : state0;
31            esac;
32
33          next(semaphore):=
34            case
35              (semaphore = semaphore0) & (state0 = enjoying): !semaphore;
36              TRUE : semaphore;
37            esac;
38
39    FAIRNESS
40      running
```

Here we define two modules: one is the main module *main* and the other is the *agent* module. The *main* module calls the *agent* module to simulate the two Agents. In the part of defining variables, lines 2 and 3 define the state space of Agent1 and Agent2, {*sleeping, trying, enjoying, exiting*}, and the fourth line defines the variable semaphore and initializes the value of the variable semaphore. FALSE, as shown in lines 9 and 10, that is, when both Agents are in the trying state at the beginning, let Agent1 enter first. The *agent* module is instantiated in the **VAR** statement, as shown in lines 6 and 7, mainly to instantiate two Agents, because to run concurrently, the keyword **process** is used. From lines 12 to line 17, we define three specifications using CTL. Line 13 checks whether the model allows both Agents to be in the enjoying state. Line 15 is to check whether Agent1 must be able to enter the enjoying state. Line 17 is to check that both Agent1 and Agent2 are in the trying state.

When the Agent module is called, the four parameters are given such as *state0*, *state1*, *semaphore*, and *semaphore0*. *state0* indicates the current state of the Agent. *state1* indicates the state of another Agent at the moment. When the Agents are in the *trying* state, in which Agents *semaphore* and *semaphore0* are equal, that Agent can be in the *enjoying* state. In the **ASSIGN** statement part, we give the Agent the initial state of *sleeping*, as shown in line 21. The case statements in lines 22–30 give the value of the next state variable *state0* of the Agent. For example, when the status of the Agent is *enjoying*, the next status can be either *enjoying* or *exiting*, as shown in line 27. The case statement in lines 32–36 gives the next value of the variable *semaphore*. The last **FAIRNESS** statement *running* is to allow the Agent process to run indefinitely.

When NuXMV is run on the program, the following output is produced:

```
1 specification EF (state_agent1 = enjoying & state_agent2
 = enjoying) is false
2 -- as demonstrated by the following execution sequence
3 Trace Description: CTL Counterexample
4 Trace Type: Counterexample
5   -> State: 1.1 <-
6       state_agent1 = sleeping
7       state_agent2 = sleeping
8       semaphore = FALSE
```

```
9 specification AG (state_agent1 = trying -> AF state_agent1
  = enjoying) is false
10   -- as demonstrated by the following execution sequence
11   Trace Description: CTL Counterexample
12   Trace Type: Counterexample
13     -> State: 2.1 <-
14       state_agent1 = sleeping
15       state_agent2 = sleeping
16       semaphore = FALSE
17     -> Input: 2.2 <-
18       _process_selector_ = ag2
19       running = FALSE
20       ag2.running = TRUE
21       ag1.running = FALSE
22     -> State: 2.2 <-
23       state_agent2 = trying
24     -> Input: 2.3 <-
25     -> State: 2.3 <-
26       state_agent2 = enjoying
27     -> Input: 2.4 <-
28       _process_selector_ = ag1
29       ag2.running = FALSE
30       ag1.running = TRUE
31     -- Loop starts here
32     -> State: 2.4 <-
33       state_agent1 = trying
34     -> Input: 2.5 <-
35       _process_selector_ = ag2
36       ag2.running = TRUE
37       ag1.running = FALSE
38     -- Loop starts here
39     -> State: 2.5 <-
40     -> Input: 2.6 <-
41       _process_selector_ = ag1
42       ag2.running = FALSE
43       ag1.running = TRUE
44     -- Loop starts here
45     -> State: 2.6 <-
46     -> Input: 2.7 <-
47       _process_selector_ = main
48       running = TRUE
49       ag1.running = FALSE
50     -> State: 2.7 <-
51   specification EX (state_agent1 = trying & EX state_agent2 =
trying) is true
```

We can note from the results of the above result of the program running. The first line illustrates that both Agents are in the *enjoying* state which is false, which means that there is only one in *enjoying* state at any time. This is what we expect. Line 51 shows that both Agents are in the *trying* state. This is true, which indicates that two Agents can compete to enter *enjoying* at the same time. It is also allowed by us, so it is also correct. From lines 9 to 51, there are counterexamples where Agent1 cannot enter the *enjoying* state when Agent1 is in the *trying* state. The reason is that on line 26, when Agent2 enters the *enjoying* state, it keeps this state forever, so Agent1 cannot enter. So we add the FAIRNESS statement to the main module, and the program is shown below.

```
1    MODULE main
2    VAR
3      state_agent1: {sleeping, trying, enjoying,
       exiting};
4      state_agent2: {sleeping, trying, enjoying,
       exiting};
5      semaphore : boolean;
6      ag1: process agent(state_agent1,
       state_agent2, semaphore, FALSE);
7      ag2: process agent(state_agent2,
       state_agent1, semaphore, TRUE);
8
9    ASSIGN
10     init(semaphore) :=  FALSE;
11
12   FAIRNESS
13     !(state_agent1 = enjoying)
14   FAIRNESS
15     !(state_agent2 = enjoying)
16
17   CTLSPEC
18     EF(( state_agent1 = enjoying) &
       (state_agent2 = enjoying))
19   CTLSPEC
20     AG(( state_agent1 = trying) ->
       AF(state_agent1 = enjoying))
21   CTLSPEC
22     EX( state_agent1 = trying) &
       EX(state_agent2 = trying)
23
24
25   MODULE  agent(state0, state1,
       semaphore, semaphore0)
26   ASSIGN
27     init(state0) := sleeping;
28     next(state0) :=
29        case
30           (state0 = sleeping)  :
```

```
            {trying, sleeping};
31          (state0 = trying) &
            (state1 = sleeping) : enjoying;
32          (state0 = trying) &
            (state1 = trying) &
33          (semaphore = semaphore0)
            : enjoying;
34          (state0 = enjoying) :
            {enjoying, exiting};
35          (state0 = exiting) :
            {sleeping};
36          TRUE : state0;
37        esac;
38
39      next(semaphore):=
40        case
41          (semaphore = semaphore0) & (state0
            = enjoying):  !semaphore;
42          TRUE : semaphore;
43        esac;
44
45   FAIRNESS
46      running
```

We added the fairness constraint given by the **FAIRNESS** statement in lines 12 and 15, so that no Agent can always be in the *enjoying* state. Let the NuXMV program run again, and the result is as follows.

```
1   specification EF (state_agent1 = enjoying &
    state_agent2 = enjoying) is false
2   -- as demonstrated by the following
    execution sequence
3   Trace Description: CTL Counterexample
4   Trace Type: Counterexample
5     -> State: 1.1 <-
6       state_agent1 = sleeping
7       state_agent2 = sleeping
8       semaphore = FALSE
9   specification AG (state_agent1 = trying ->
    AF state_agent1 = enjoying)  is true
10  -- specification (EX state_agent1 =
    trying & EX state_agent2 = trying)  is true
```

From line 9, we can see that $\mathbf{AG}((state_agent1 = trying)-> \mathbf{AF}(state_agent1 = enjoying))$ is now satisfied.

References

1. V. Aho, J.E. Hopcroft, J.D. Ullman, *The Design and Analysis of Computer Algorithms* (Addison-Wesley Longman Publishing Co., Inc., Boston, 1974)
2. R. Alur, T. Henzinger, Reactive modules. Form. Methods Syst. Des. **15**, 7–48 (1999)
3. D. Broman, E. Lee, S. Tripakis, M. Törngren, Viewpoints, formalisms, languages, and tools for cyber-physical systems, in *6th International Workshop on Multi-paradigm Modeling (MPM'12)* (2012)
4. J.R. Burch, E.M. Clarke, K.L. McMillan, D.L. Dill, L.J. Hwang, Symbolic model checking: 10^{20} states and beyond. Inf. Comput. **98**(2), 142–170 (1992)
5. R.E. Bryant, Graph-based algorithms for boolean function manipulation. IEEE Trans. Comput. **C-35**(8), 677–691 (1986)
6. A. Cimatti, E.M. Clarke, E. Giunchiglia, F. Giunchiglia, M. Pistore, M. Roveri, R. Sebastiani, A. Tacchella, NuSMV 2: an open source tool for symbolic model checking, in *CAV*, ed. by E. Brinksma, K.G. Larsen. LNCS, vol. 2404 (Springer, 2002), pp. 359–364
7. E.M. Clarke, E.A. Emerson, Design and synthesis of synchronization skeletons using branching time temporal logic, in *Logic of Programs: Workshop*, Yorktown Heights, NY, May 1981. LNCS, vol. 131 (Springer, 1981)
8. E.W. Dijkstra, The humble programmer. Commun. ACM **15**(10), 859–866 (1972)
9. E.M. Clarke Jr., Orna Grumberg, Lucent Technologies, *Model Checking* (MIT Press, Cambridge, MA, 1999)
10. E.A. Emerson, E.M. Clarke, Characterizing correctness properties of parallel programs using fixpoints, in *Automata, Languages and Programming*. LNCS, vol. 85 (Springer, 1980), pp. 169–181
11. E.A. Emerson, J.Y. Halpern, "Sometimes" and "Not Never" revisited: on branching time versus linear time. J. ACM **33**, 151–178 (1986)
12. E.A. Emerson, C.-L. Lei, Efficient model checking in fragments of the propositional mu-calculus, in *LICS86* (1986), pp. 267–278
13. W. Fokkink, *Introduction to Process Algebra* (Springer, Heidelberg, 2000)
14. N. Francez, *Fairness* (Springer, 1986)
15. M.R. Garey, D.S. Jolmson, *Computers and Intractability: A Guide to the Theory of NP-Completeness* (W. H. Freeman and Company, San Francisco, 1979)
16. C. Hoare, *Communicating Sequential Processes* (Prentice Hall, New York, 1985)
17. J. Hopcroft, R. Motwani, J. Ullman, *Introduction to Automata Theory, Languages, and Computation*, 3rd edn. (Addison-Wesley, Reading, 2006)
18. G.E. Hughes, M.J. Creswell, *Introduction to Modal Logic* (Methuen and Co. Ltd., London, 1968/1977)
19. Z. Manna, A. Pnueli, *Temporal Verifications of Reactive Systems-Safety* (US, Springer, New York, 1995)
20. K.L. McMillan, *Symbolic Model Checking* (Kluwer Academic Publishers, Norwell, 1993)
21. W. Reisig, *Petri Nets: An Introduction* (Springer, Heidelberg, 1985)
22. https://es.fbk.eu/tools/nuxmv/downloads/nuxmv-user-manual.pdf

Chapter 4
QoS for 5G Mobile Services Based on Intelligent Multi-access Edge Computing

Stojan Kitanov, Tomislav Shuminoski, and Toni Janevski

4.1 Introduction

The edge (mobile, or fixed) terminal equipment in the future broadband networks and services would be required to collect a large amount of data in a reliable way as well as to provide valuable data to the cloud computing for big data analytics. Also, it should include machine learning (ML), advanced machine-to-machine communications, and intelligent edge computing with artificial intelligence (AI) within, in order to have the capability to analyze the data and to respond promptly with a high level of QoS to the end users [1].

The cloud computing and QoS mechanisms are expected to be diffused among all network entities, including the smart mobile edge devices in 5G and future mobile networks, i.e., the cloud computing would be extended at the network edge in a form of intelligent Multi-access Edge Computing. Like that geo-distributed edge nodes and user's mobile devices would cooperate with each other and with the cloud computing data centers, in order to distribute data computation and processing and network services near to the consumers [2].

Mobile and fixed broadband refers to the bitrates that give seamless access to all existing web applications and services. However, ultra-broadband is a term which is used by many national governmental strategies for broadband in 2020 and beyond in order to refer to higher bitrates than the currently available broadband [3].

S. Kitanov (✉)
Faculty of Informatics, Mother Teresa University, Skopje, Macedonia
e-mail: stojan.kitanov@unt.edu.mk

T. Shuminoski · T. Janevski
Faculty of Electrical Engineering and Information Technologies, Ss Cyril and Methodius University, Skopje, Macedonia
e-mail: tomish@feit.ukim.edu.mk; tonij@feit.ukim.edu.mk

© Springer Nature Switzerland AG 2021
H. Gao, Y. Yin (eds.), *Intelligent Mobile Service Computing*, EAI/Springer Innovations in Communication and Computing,
https://doi.org/10.1007/978-3-030-50184-6_4

Nowadays, the existing broadband access is in range of Mbit/s per user to tens of Mbit/s per user. On the other hand, the ultra-broadband access aims for bitrates in the range from several hundreds of Mbit/s per user up to Gbit/s per user.

The QoS in 5G network and the mobile cloud computing mechanisms deliver both mobile broadband access systems and services which are optimized for high spectral efficiency, high throughputs, energy efficiency, high level of reliability, and lower network delays, with the following advanced technologies: link adaptation, machine-to-machine communications, fast handoff, fast power control, superuser MIMO techniques, etc. [4]. Moreover, it would enable direct communication among the devices and would provide a support of the network function virtualization, as well as the segregation of intelligent network control from the hardware in the radio network. Like that network resources would be utilized in the best efficient manner. In addition, QoS mechanisms, cloud and mobile edge computing mechanisms would be able more efficiently and effectively to deal with a huge amount of big data processing and more intelligent networking demands, such as reduced network delay, a high level of seamless mobility, high adaptive scalability, and running the applications in the real time.

Compared with the past network deployments, the 5G mobile systems demand devices capable to supply the consumers with mobile broadband services, omnipresent versatility, capabilities for advanced mobile cloud computing mechanisms, intelligent edge computing features, large processing capability of mobile terminals, communications among the machines without any human interference, improved utilization of the network, etc. However, the main goal is to provide a support for high QoS level, as well as faster data processing features and improved life of the battery in the mobile devices.

In 5G mobile network and beyond, each mobile device would use different data and services from many various radio access technologies, providing a high level of QoS.

This chapter is about 5G mobile networks and their cloud computing and QoS mechanisms. Furthermore, the advanced QoS for 5G mobile services based on Intelligent Multi-access Edge Computing together with the capability of radio network aggregation is presented. The framework of 5G leads to a provision of high QoS for any multimedia service, improved utilization of the transmission bandwidth and load sharing of the traffic, and improved features of mobile cloud and fog computing and capabilities for multiple radio interfaces on the device. The QoS and cloud-based framework approach are user-centric, focused on continuously online network by using the radio network aggregation of available mobile broadband connections enabled with the focal points of versatile mobile cloud and fog computing. Finally, an overview of both the vertical multi-homing and the multi-streaming features for mobile and fixed terminals together with the capability of radio network aggregation in 5G networks is elaborated.

4.2 5G QoS

These days, the world of ICT is in exceptionally quick huge development and improvement. This is especially true about the technologies used in 5G network. 5G would provide an endless extend of services to consumers, with omnipresent versatility, big processing capability of the mobile terminal, support of advanced QoS, and systems and networks being optimal for high spectral efficiency, high throughputs, energy efficiency, high level of reliability, lower network delays, utilizing progressed advances of technologies such as link adaptation, machine-to-machine communications, fast handoff, fast power control, Smart Home with IoT (Internet of Things) platforms and many other advanced capabilities [5].

The use of both smart mobile wireless devices and their applications is now a necessary portion of everyone's lifestyle. Mobile devices have increasing processing capabilities, so in many cases, they have supplanted user's fixed landline telephony, video camera, personal computer, and even television. At the present, millions of videos are viewed online, and thousands of images are uploaded or downloaded every minute.

At the moment the Internet progressively evolves from Internet of Computers into *Internet of Things* (IoT) and moreover into *Web of Things* (WoT) [6]. In the future the Internet would clearly evolve from the Internet of Things (IoT) into the *Internet of Everything* (IoE). This would introduce raise new challenges on the services that would be provided to the consumers, such as high level of mobility, high scalability, and requirements to run the applications and services in real time and to execute them with a low latency [7].

IoT could be a framework that interconnect intelligent physical objects or "things" that possess hardware, software, sensors, actuators, and network connectivity which empowers these objects to gather and transmit data through various networking interfaces and Internet [6]. IoT empowers objects to be detected and/or controlled remotely through the existing network infrastructure. The smart things may contain their personal IP addresses, can be inserted in complex systems, utilize sensors to gather information about the environment, and use actuators for environment's interaction. If IoT is expanded with sensors and actuators, then it becomes an instance of cyber-physical systems' class that encompasses smart cities, intelligent transportation systems, smart homes, and smart grids.

WoT integrates IoT on both the Internet (the network layer) and the web (the application layer) [7]. Like that the creation of IoT applications is simplified. The physical devices can be accessed as web resources by using standard web protocols, where the services/applications are provided with a web-based service environment or legacy telecommunication networks. Therefore, the integration of the systems and applications is easier. The services and data offered by the smart web things would be available to all application developers, from which they can build new innovative, interactive, and scalable applications which would be beneficial for everyone.

Internet of Everything (IoE) would represent an intelligent connection of the following entities people, process, data, and things [7]. These entities are brought together in order to establish more valuable and relevant connections than ever before. Like that information is transformed into an action that can create new richer experiences and capabilities and new economic opportunities for countries, businesses, and individuals. The main drivers which would enable IoE to become reality are the development of IP capable devices, the worldwide availability of mobile broadband services, and the invention of IPv6. In addition, 5G network would have a great impact on IoE, because it would provide a necessary secure infrastructure which is intelligent, manageable, and scalable to provide a support of billions of devices aware of the context.

However, there was an operator-centric approach used in 3G mobile networks such as UMTS/HSPA, cdma2000, Mobile WiMAX 1.0 (IEEE 802.16e). On the other hand, a service-centric approach was deployed in 4G mobile networks; also referred to as IMT-Advanced umbrella are the following technologies: LTE Advanced and Mobile WiMAX 2.0 (802.16 m), [8, 9]. 5G mobile network shifts toward the user-centric concept, like LTE Advanced Pro, IEEE 802.11 ac/ad/af, and others, also known as IMT-2020 umbrella [9–16]. In the future, every mobile terminal with their multiple interfaces would have access to different wireless technologies at the same time, and it would combine diverse streams from various Radio Access Technologies (RATs) using advanced QoS algorithms, control protocols, and mobile cloud computing mechanisms.

On the other side, the existing mobile and wireless networks are already migrating toward single all IP-based network. In other words, all data and signaling traffic would be transmitted via the standard Internet Protocol on the network layer. In spite of the fact that distinctive RATs, such as 3G UMTS, Mobile WiMAX, LTE, LTE Advanced, and LTE Advanced Pro, may co-exist together, the common "thing" which unifies all of them is Internet Protocol (IP). That's why the most alteration changes and advancements are expected to be performed in the network layer of the mobile terminal.

Furthermore, each wireless network is responsible for overseeing user mobility, whereas the ultimate choice among diverse wireless and mobile access network providers (different RATs) for a given multimedia service, the mobile terminal, and the edge computing entity. In that context, the provision of the QoS in the networks has become a very important objective, because it demands great thoughtfulness, extraordinary versatility, and profound examination.

One of the foremost valuable assets in remote and portable frameworks is the radio bandwidth. Therefore, an efficient QoS framework is needed to guarantee the necessary QoS and simultaneously to perform a maximal utilization of the bandwidth in all available mobile and wireless and networks.

On the other hand, there exist different multimedia traffic flows in heterogeneous wireless and mobile network and each of them with different QoS requirements. Therefore, it is necessary to achieve an increased access probability for any particular service. In that manner, having a bandwidth scarcity, the increasing of the throughput and keeping up the high value of multimedia access probability at the same time are an awesome advancement due to the reality that the system throughput and the access probability are clashing with each other.

4.2.1   5G Features

5G represents a set of evolved and enhanced existing and novel network technologies. However, 5G has a target to secure unlimited access to data and information sharing at any place, anytime by anyone, or anything for the wellness of society, business, and people. The performance levels and requirements that technologies and systems that are necessary to meet in order themselves to be marked as 5G are already standardized. The standardization that started in the past years leads to commercial availability of both the equipment and the devices in 2020, the year marked as 5G. All ICT players collaborate with each other in order to evolve and upgrade existing technologies which would secure a smooth transition from 4G to 5G.

However, 5G represents much more than upgraded set of existing technologies. It would bring together both the advanced forms of as of now accessible multi-radio access, core technologies, and cloud computing technologies together and the new complementary technologies, in order to handle much more traffic and provide a support for more devices with different operating requirements, regulations, and many different use cases. Compared with the Fourth Generation (4G) of mobile and wireless technologies, the estimated levels of performance that 5G would have to cater for are the following [17]:

- Increased traffic volume by a factor of 1000
- Tens of billions of connected devices
- Improved achievable user data rates by a factor ranging from 10 to 100
- Reduction of the delay up to a factor of 10
- Improvement of the battery life by a factor of 10
- Support for various IoT devices such as mobiles, tablets, embedded devices, etc.
- Security and integrity of data

The many advantages of 5G, as Fig. 4.1 presents, could have a significant impact in the technological development, as well as the economic and social development in any country. Through the digitalization of the society, 5G is anticipated to contribute to countries' financial development.

The specifications for 5G, as well as for next generation of mobile telecommunications, are set primarily by following bodies: International Telecommunication Union (ITU) and 3rd Generation Partnership Project (3GPP). According to the ITU, the initial 5G commercial deployments would appear around 2020. The exploratory stage that started with a reason to determine the demands and identify the foremost promising techniques and technologies for 5G networks is about to be completed. ITU plays an imperative part in characterizing the technologies and standards that govern any new generation of IMT globally. The IMT standards, which have been defined with the association of numerous public authorities and industrial players, have established the framework for the evolution of mobile communication services around the world, since the very beginning of IMT standardization, starting

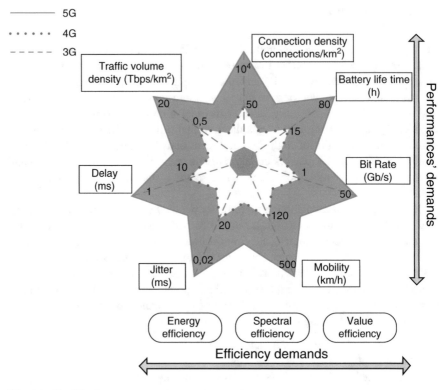

Fig. 4.1 The 5G advantages

with IMT-2000 (3G, UMTS), then IMT-Advanced (4G, LTE Advanced), and most recently IMT-2020 (5G, LTE Advanced Pro).

ITU, under IMT-2020 and 5G Infrastructure Public-Private Partnership (5G-PPP), has defined the following three main use cases, with their respective – and possibly commonly inconsistent – requests that are within the process of taking shape and will make it conceivable to meet the sector-specific needs [17]:

- Massive Machine Type Communications (mMTC): represent a really huge number of connected devices with disparate quality of service requirements. The primary objective is to supply a incite reaction to the exponential increment within the density of connected objects.
- Enhanced Mobile Broadband (eMBB): Giga speed indoor and outdoor connection, with uniform QoS, including network's edge.
- Ultra-Reliable and Low Latency Communications (uRLLC) or Critical Communications: requires very strict demands for latency and packet loss, in order to guarantee expanded reactivity.

The mMTC essentially envelops all IoT related utilizations, where the services require broad coverage, lower energy consumption, and relatively slow transmission speeds. Compared to the existing technologies, 5G would make it possible to connect objects that are densely geo-distributed across a given area.

Enhanced mobile broadband (eMBB) is primarily concerned with the applications and services which require fast connections, for example, to watch ultra-high-definition (8 K) videos or to stream wirelessly virtual and augmented reality applications.

Ultra-reliable and low latency communications (uRLLC) contain the applications which require a prompt response and very strong ensures of message transmission. These requirements prevail particularly in the transportation sector (e.g., the response time when an accident occurs) and medicine (telesurgery) and when the fabricated processes are digitized in general.

In the near future, not only people and systems would create and share data, but billions of IoT devices would also be a significant portion of the communications framework. This framework would envelop cloud computing, Multi-access Edge Computing, IoT platforms, virtualization (SDN and NFV), as well as more and new devices and new business models.

In order to meet all of the above service requirements, 5G networks would face with significant design challenges. When 5G would be commercially deployed, they must meet a number of individual users and enterprise needs [18]:

- Immersive experience – Minimum 1Gbps or up to 50 Gbps data rates in order to provide a support of ultra-high-definition video and virtual reality applications
- Fiber-like user experience – Data rates in the order of 10 Gbps to support mobile cloud computing services
- Zero-second delay and response – Less than 1 ms delay which would provide a support for mobile control and vehicle-to-vehicle (V2V) communications in real time
- Zero-second switching – Maximum 1 ms switching time between different radio access technologies in order to secure a permanent and seamless delivery of services
- Massive capacity and always-on operation –Support of several billions of applications and hundreds of billions of devices
- Energy consumption – Reduction of energy-per-bit consumption by a factor of 1000 in order to extend the battery life of the connected device

Overall, 5G mobile network has a user-centric concept, as our proposed 5G concept in this chapter [9]. One may expect that the mobile and fixed terminals simultaneously to be connected to diverse mobile and wireless technologies, as well as to be able to combine different traffic streams from various access technologies in order to obtain very high bitrates (e.g., 10 times higher for individual users than those in 4G and 1000 times higher aggregate bitrates). Therefore, the 5G represents a fundamental shift in the wireless and mobile networking philosophy toward user-centric mobile broadband networks with a high level of QoS provision and Intelligent Multi-access Edge Computing.

4.2.2 QoS

Today there are a number of definitions about the term quality of service (QoS), depending on the perspective of viewing, ranking, and evaluating both the end user and the network side, on a certain set of connection parameters that should guarantee a satisfactory level for a given service (application) which is offered by the network.

ISO 8402 characterizes QoS as: "a set of characteristics of a particular entity that contains the ability to meet certain specific needs (requirements)." On the other hand, ISO 9000 defines quality as a level or degree in which certain features meet their prerequisites. The International Telecommunications Union ITU-T (ITU-T Recommendation E800) and ETSI (Recommendation ETSI-ETR003) define QoS as: "the collective effect of service performance, which determines the satisfactory user level for that specific service." Furthermore, ETSI has developed another definition (according to ETSI-TR102) where QoS is presented as the capability to segment traffic or the capability to differentiate between different traffic types, and therefore the network itself can treat each traffic stream differently. This definition clearly specifies the importance of differentiating streams within the network so that different QoS mechanisms can be used on different streams and treated differently (e.g., for real-time and off-time services, one requiring minimal delay and jitter, and to others impeccable accuracy during transmission).

Despite all these definitions, one key point is that, from a network perspective, QoS is the ability of network elements (routers, mobile nodes, hosts, servers, etc.) to have some level of security that they can meet all traffic and service requirements. To emphasize, the main idea is not a flawless level of QoS but rather a satisfactory level of quality assurance of the given service within the network or the networks in which the service is delivered.

It is quite clear that the QoS mechanisms themselves in wireless and mobile systems have a huge challenge to deal with a huge number of constraints, unlike the mechanisms used in fixed networks. Thus, QoS mechanisms in wireless and mobile systems are closely related to bandwidth management (spectrum is one of the most expensive resources), access, number of users, power consumption, and network management settings depending on the application (service) requirements. It is obvious that all applications (whether real or non-real time) require satisfying certain QoS parameters. Generally, all services can be treated in one of the six QoS classes given in Table 2/Y.1541, according to ITU-T Recommendation Y.1541 [19]. Moreover, in [19] (Table 1/Y.1541) the main QoS requirements and provision of IP network QoS class definitions, as well as the network performance objectives, are provided. Consequently, for each particular class of service, there are specific QoS specific parameters that must be observed, and on the other hand, some of the parameters do not even have to be specific to certain applications (services).

According to the above, generally speaking, QoS means changes, both in terms of application and scientific area. Thus, there are generally three types of QoS in heterogeneous mobile and wireless networks: *essential QoS, perceived QoS, and fixed QoS*.

Essential QoS can be directly measured and controlled by the telecommunications network itself and in many cases described by objective QoS parameters (such as packet loss, delay, delay variation, etc.). The usual QoS (perceived QoS) type of QoS is still referred to as QoE (Quality of Experience) and is the quality that users perceive.

QoE is closely related to network performance; however, it is evaluated by the "middle ground" of the users, i.e., MOS (Mean Opinion Score). This method is used to measure perceived QoS by the user and is represented by the following ratings: 1, bad quality; 2, poor quality; 3, fair or satisfactory quality; 4, good quality; and 5, excellent quality. In doing so, MOS is the arithmetic mean of all ratings given by surveyed users in the range of 1 to 5.

However, to emphasize that although there is a close relationship between the objective network metrics and user perspective on quality, users can never accurately evaluate them. Substantial QoS variations are caused by objective performance reductions/increases in service quality. Although these two terms QoE and QoS are often overlapping in the literature and are confusing from a service point of view, one thing is clear: QoE represents the entire network implementation of quality from the user's perspective and is also a subjective measure of performance. QoE is an excellent indicator of how networking opportunities meet customer needs and expectations for the quality of a given service. QoS, on the other hand, as we have already pointed out, is a performance measurement at the packet-level from a network point of view, containing a set of mechanisms and objective QoS parameters that allow the network administrator to implement different QoS for different traffic flow for a given user. Finally, the set of QoS parameters consist of both the essential QoS and the perceived QoS, i.e. initially, the desired core QoS is tested, and then it is compared with the results obtained from the users, i.e., perceived QoS.

Moreover, in each network QoS class makes a particular combination of bounds on the performance values. Any flow that satisfies all the performance objectives of a QoS class is totally compliant with the normative recommendations of ITU Recommendation for that class.

Table 4.1 provides some guidelines for the application and engineering of the network QoS classes. It also conveys one of the principles of QoS class development that the requirements of multiple applications are addressed by a single set of network performance objectives. This approach keeps the number of QoS classes small and manageable.

In achieving end-to-end QoS, some of the fundamental challenges are present when:

- Many network providers should complete the path.
- The number of networks in the path would vary from request by request.
- Distances among the users are generally unknown.
- There are high variations in the impairment level of any given network segment.
- It is desired to evaluate the actual performance levels achieved on a path.

Table 4.1 Guidance for IP QoS classes [20] (ITU-T Y.1542)

QoS class	Example of service	Node mechanism	Network technology
0	Real-time, jitter-sensitive, highly interactive (VoIP, VTC)	Separate queue with preferential servicing, traffic grooming	Constrained routing and distance
1	Real-time, jitter-sensitive, interactive (VoIP, VTC)	Separate queue with preferential servicing, traffic grooming	Less constrained routing and distances
2	Transaction data, highly interactive (signaling)	Separate queue, drop priority	Constrained routing and distance
3	Transaction data, interactive	Separate queue, drop priority	Less constrained routing and distances
4	Low loss only (short transactions, bulk data, video streaming)	Long queue, drop priority	Any route/path
5	Traditional applications of default IP networks	Separate queue (lowest priority)	Any route/path

- The operator should be capable to tell whether the requested performance can be met or not.
- The process must eventually be automated.

4.2.3 5G Network Slicing

As it was mention in Sect. 4.2.1 many applications such as mMTC, eMBB, uRLLC, and Critical Communications shall simultaneously share and run the 5G physical network [17]. These applications demand various QoS requirements and transmission bandwidth. For example, the eMBB video-on-demand streaming applications demand very high transmission bandwidth and transmit a big quantity of data. On the other hand, mMTC applications, for example, IoT, require low throughput for the transmission.

The one-size-fits-all approach of the conventional systems does not satisfy these various requirements and demands for all these vertical services. One possible cost-effective and cost-efficient solution is to divide or slice the physical network into many isolated logical networks, i.e., to use **network slicing**. Network slicing represents virtualization in which the shared physical network infrastructure is partitioned into many end-to-end level logical networks that would allow the tenants to be grouped or isolated by the type of traffic. Like that the network slices can be seen as a group of virtual networks on the top of a common underlying physical infrastructure [21].

Network slicing actually practically implements the network as a service (NaaS) model. This enables the service providers to create and set up their own network infrastructure according to their demands and to perform a customization for different and sophisticated scenarios. Therefore, network slicing is one of the main enabling technologies in 5G network and beyond.

Some of the network slicing building blocks that support network virtualization are software-defined networking (SDN) and network function virtualization (NFV). SDN separates the tightly coupled control and data planes of traditional networking devices and provides a central view of the network. Like that it is a possible to have a single point of network administration from where run network control functions can be performed. NFV, on the other hand, performs a transfer of the network functions from exclusive equipment hardware- to software-based applications which are performed on general-purpose hardware. Therefore, the cost is reduced, and the elasticity of network functions is increased by creating connected or chained virtual network functions (VNFs) in order to build communication services.

Network slicing uses network resources in the most cost-effective way and reduces both CAPEX and OPEX [22]. Like that reliability and limitations such as congestion and safety aspects of one slice do not affect the other slices. Moreover, the isolation and security of information, control, and management plane are also assisted with network slicing, and like that network security is provided. Finally, network slicing can be expanded to cloud computing and mobile edge computing that eventually improves their interoperability and helps to bring services closer to the end user with fewer violations of service-level agreement (SLA) [23].

However, there are many challenges in network slicing that need to be resolved in order to be fully implemented in the 5G network and beyond. For example, every virtual network demands a different level of resource affinity, and it can be changed with the course of time. Therefore, it is difficult to achieve resource provisioning among multiple virtual networks. In addition, end-to-end slice management and orchestration, as well as mobility management and wireless resource virtualization, can increase the network slicing problems in 5G.

Various researches address these challenges through an efficient 5G network slicing framework. This framework contains the following three layers: infrastructure layer, network function layer, and service layer, as it is illustrated in Fig. 4.2 [24, 25].

The *infrastructure* layer characterizes the actual physical design, which incorporates Edge Cloud and distant Cloud Information Center. Here various software defined techniques provide deliberation of the resources. In addition, several policies are conducted to deploy, control, manage, and orchestrate the fundamental infrastructure. Moreover, a few approaches are conducted to send, control, oversee, and coordinate the fundamental infrastructure. This layer gives transmission capacity, storage, and computing assets to network slices in such a way the higher layers are able to access and manage them.

The *network function and virtualization* layer performs all the necessary operations that are needed to handle the virtual assets, as well as network function's life cycle. Throughout the usage of SDN and NFV, it provides support the network slices to be optimally placed in order to meet the service and application requirements. In addition, this layer handles the functionality of local radio access and core networks

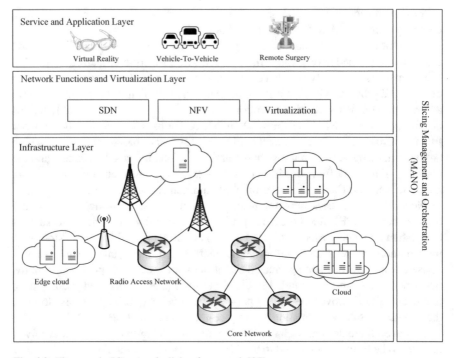

Fig. 4.2 The generic 5G network slicing framework [25]

in an explicit way. It also deals with fine-grained and coarse-grained network functions in an efficient way.

The *service and application layer* consists of connected smart mobile devices, virtual and augmented applications, etc. which have a particular usage. The VNFs are mapped to physical assets according to the service and application requirements in the service-level agreement.

These layers are overseen and handled by the slicing management and orchestration layer (MANO). Some of the main functionalities of this layer are the following:

- Assisting the infrastructure layer to creating virtual network instances upon the physical network
- Helping the network function and virtualization layer to make a service chain by performing a mapping between the functions of the network to virtual instances of the network instances
- Handling the scaling of the virtual assets in accordance to the context changes by maintaining the communication interface between service and application and the network functions and virtualization layer

However, yet the development the network slicing framework in 5G is in progress. This may be a solution for 5G network and beyond, to extend the network slicing framework to deal with the dynamics of future virtual assets.

4.3   Mobile Edge and Cloud Computing Architecture and Design

The Intelligent Multi-access Edge Computing (IMAEC) is a new concept and network architecture that enables advanced edge networking, intelligent data processing capabilities, and QoS provisioning for all used multimedia services by applying ICT with QoS mechanisms and optimization algorithms.

As can be noticed in [26], there are two main aspects: autonomous network control for an intelligent edge networking capability and ambient intelligence analytics for intelligent data processing capability. Figure 4.3 shows that one IMAEC may consist of several important entities (deployed in physical or virtual nodes or equipment).

The main IMAEC entities are:

- RATs from different technologies (2G, 3G, 4G, 5G, and beyond), as access nodes (base stations or access points to the mobile equipment).
- Edge DB is database within the Intelligent Multi-access Edge Computing system, which stores all information, parameters, and variables needed in this system.

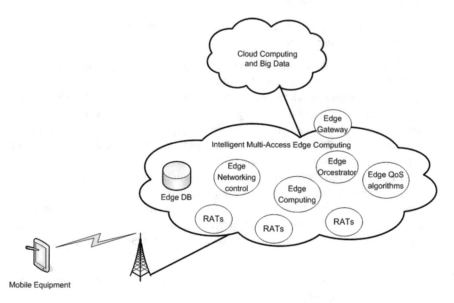

Fig. 4.3 Illustration of the intelligent edge and cloud computing entities

- Edge networking control (ENC), which is providing connectivity, routing, and traffic control to mobile equipment, using heterogeneous mobile and wireless technologies (RATs). Moreover, it is responsible for the optimal usage of the available resources in the constrained heterogeneous environment. This entity is between the mobile equipment and the edge computing entity.
- Intelligent computing entity (ICE), which provides edge analytics of AI service on itself or other analytics, such as big data analytics on cloud computing. Thus, it performs data analysis functions based on gathered information and doing the computation of the requirements asked (or required) from mobile users and services.
- Edge orchestrator, which orchestrates the edge computing processes in an autonomous way. The orchestrator is doing the automated configuration, coordination, and management of Intelligent Multi-access Edge Computing systems and servers. Also, it provides a management function that stores the identity of entities such as mobile equipment, IoT devices, cloud servers, resources in each server, and routers in some areas, which includes data names as an identity. In addition, it performs mapping of the identities to metadata locations. It could, therefore, be involved in mobility management functions.
- Edge QoS algorithms, which store the most frequently used algorithms for optimization and other processes important for the QoS mechanisms. So, this entity is very crucial for the QoS and cloud provisioning.
- Edge gateway entity (EGE) is an entity that provides an interworking function to outside entities including other Intelligent Multi-access Edge Computing systems and cloud computing with big data analytics. Thus, it can function as a gateway function or border intelligent router.

One Intelligent Multi-access Edge Computing system should consider the following two important capabilities:

- Intelligent edge networking and routing capability: In order to support mobility, real-time communication, reliable communication, scalability, high QoS provisioning, constrained environment, available spectrum, and easy deployment, the IMAEC should consider networking technologies, such as applying various wireless and mobile technologies (RATs) and composing networking dynamically. Moreover, it should include backward compatibility with existing networks and forward compatibility with the future networks.
- Intelligent data processing and computing capability: In order to provide support of edge analytics, IMAEC has to increase data utilization by using intelligent data processing, data collection, and dynamic storage, using adequate optimization algorithms and real-time trust data process. Especially, analysis models should be updated periodically by some big data servers in the cloud computing network or center.

More details about the signaling requirements, protocols, and architecture for intelligent edge computing can be found in [26].

Furthermore, Fig. 4.4 presents the general architecture of the intelligent edge and cloud computing for using mobile cloud services by the consumers. The 5G mobile equipment (ME), the Intelligent Multi-access Edge and cloud computing servers are equipped with vertical multi-homing and multi-streaming features and resource optimization algorithms that are able to execute a simultaneous handling of multiple radio network connections as well as to enable a faster transfer of various multimedia services. That is done through transferring each object of each service through a particular stream or sub-streams, to obtain the best level of end-to-end QoS. On the other hand, the edge computing and cloud computing are directly connected to the cloud server in the middle of the core part of the network, and part of the jobs can be transferred or balanced with the cloud server. Also, the cloud server over the Transit GW is connecting to big data and public cloud computing servers but also is connected to the multimedia servers with assured QoS provisioning. So rather than to create an isolated connection for every object

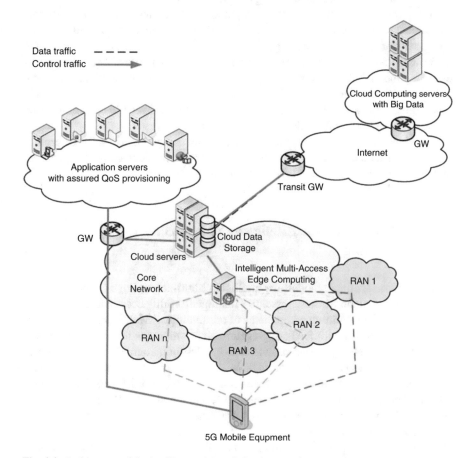

Fig. 4.4 Architecture of the intelligent edge and cloud computing

as in TCP, the intelligent cloud and edge computing use its own multi-streaming and multi-homing feature to increase the transfer bit rate of the target service over isolated streams through many different RATs. However, all multimedia services which travel over the core networks and any available 3G, 4G, or 5G RATs terminate in the user's 5G mobile terminal equipment in downstream and upstream direction.

The cloud server in the core network is a high-performance data computing center with massive processing power, and it is able to perform processing and analytics on big data that cannot be performed on edge computing devices. It contains the cloud orchestrator and the cloud database.

The cloud orchestrator provides orchestration mechanisms which ensure resilient and trustworthy open, dynamic, and large scale of services [27]. It is responsible for composing the available service elements in the edge computing environment, which includes to sense, connect, store, and process data via platform and software services, into more complex mobile edge computing services that can be offered to the users, such as traffic crowdsensing and trip planning services.

Executing of the mobile edge computing services involves many different components and entities that are geo-distributed in a large area, and thus the complexity of the decision-making process in what regards the resource allocation to attain acceptable QoS/QoE levels is increased. In order to perform a coordination of executing the mobile edge computing services, the orchestrating mechanisms ought to be able synchronize and combine the operation of the various service elements in order to meet the necessary requirements of the composed mobile edge computing services, about low delay, scalability, and resilience.

The cloud database contains the profiles of cloud computing users. These profiles contain information about each user's subscriptions to the cloud computing services and determine the user access level to cloud computing services.

4.4   System Model for Intelligent Multi-access Edge Computing

Although there are many models for Intelligent Multi-access Edge Computing platforms, here are elaborated one essential Intelligent Multi-access Edge Computing system model. The building components of our Intelligent Multi-access Edge Computing system model with RATs aggregation algorithm for radio network selection and routing in heterogeneous mobile and wireless environment are provided in Fig. 4.5. Data measurements for various criterions of selection, which include user requirements, essential QoS requirements, different RAT links' conditions, edge computing requirements and conditions, and cloud computing requirements, represent the input for the n sets of parallel criteria functions (CFs), one set per each RAT (from RAT 1 up to RAT n).

A single RAT CF performs shaping and filtering on the outputs from the previous (five) components into four interior threshold functions: the first shapes the QoS

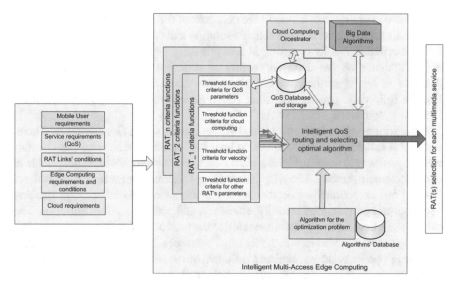

Fig. 4.5 Functions of the Intelligent Multi-access Edge Computing

parameters, the second shapes the cloud computing resources and conditions, the third performs shaping of velocity support, and the last defines the RAT parameters (mainly the strength of the signal) that are detected in the mobile terminal from RAT base station(s). Each of those four CFs provides only one value on its output.

The processes of computing, filtering, and shaping are executed in the Intelligent Multi-access Edge Computing Server (which is the most appropriate) or in the mobile device. This depends primarily on the availability of the measured values, location of measurements, available computing resources, and complexity. For example, the threshold function for battery support and internal memory of mobile equipment, which is shaping and filtering the battery level parameter and memory (in GB) within the mobile equipment, should be placed in the mobile equipment. On the other side, the threshold functions for RAT's parameters and cloud computing requirements, which are shaping and filtering the service QoS and intelligent edge computing features, should be placed in the Intelligent Multi-access Edge Computing platform (servers).

The intelligent QoS routing and selecting optimal algorithm module is the central component. It has ability to choose one most appropriate optimization algorithm. The outputs of the n sets of parallel criteria functions (CFs), i.e., four values from each RAT (4*n in total) are used as inputs. The selection of the most appropriate algorithm for the optimization problem is defined by the given scenario and available RATs. It depends on the various input criteria for each RAT for a given multimedia service from the algorithm's database. In order to provide an optimal solution, this module is orchestrated by the cloud computing orchestrator to do the optimal solution. Besides those entities there are QoS database and storage, and if this module for intelligent QoS routing and selecting optimal RAT(s) is placed in the cloud computing data center in the core part of the network, it can contact also algorithms for big data for more reliable solution and optimization.

Each criterion has its own weight that primarily depends upon the assumption of its influence on the decision process of selecting the best RAT. Within the database in the algorithm, we are proposing stored knowledge and algorithms like the following: genetic algorithm (GA), linear programming (LP) algorithm, Lyapunov algorithms, etc.

LP algorithm gives better yields and is less complex than GA algorithm. Therefore, LP is suitable to the solution of optimal RAT decision problems. The GA may well be pertinent to more straightforward issues and capacities with small trouble in computational overhead. The adaptive queuing Lyapunov optimization procedures [28–31] are exceptionally effective for the optimization of time average queuing networks and provide joint stability, efficiency, and performance optimization. The existing solutions and application maximize the network throughput that is subject to average power constraints on the terminal interfaces. The solutions also minimize the average queue backlogs, subject to minimal queue network delay and in the same time achieve stability of the network, as well as high spectral efficiency. Moreover, the Lyapunov optimization refers to the use of a Lyapunov function and Lyapunov stability in the optimally control and stability of the dynamical system.

In general, although the complexity of this framework is significantly reduced in with the proposed algorithms, due to the usage of such output ranking functions; however, it can be extended to novel and more complex algorithms and optimization problems.

The QoS database stores information about QoS parameters for all services and measured data and provides an additional storage space for the cloud computing processes. Moreover, in the database with big data algorithms are stored optimization algorithms that are used to solve concrete problems for optimization in big data analysis and deep learning.

Finally, the intelligent QoS routing and selecting an optimal algorithm module (together with the other entities around this module), besides other already mentioned functions, aims to select the most appropriate RATs in a heterogeneous environment. Therefore, the decision process should select the best RAT or RATs from all present RATs for a given user and given service and would send the multimedia stream through multi-homing and multi-streaming features [32, 33].

Furthermore, Fig. 4.6 presents Intelligent Multi-access Edge Computing system model. As can be seen, our proposed system model for 5G mobile terminal (mobile part) and for the Multi-access Edge Computing server (fixed part) is multi-access equipment with Intelligent Multi-access Edge Computing support. It contains several RAT interfaces, each for a particular RAT (for all available RATs). The intelligent multi-access end user computing module, which has features of the QoS routing and selecting optimal algorithm, is located in the IP layer. According to [10, 14] physical Open Wireless Architecture (OWA) layer determines the wireless technology. As can be noticed, the OWA layer can be added to any present or future RAT with their well-defined medium access control (MAC) and physical layers, because all of them nowadays have IP as a network protocol layer, above the MAC layer (OSI-2 layer).

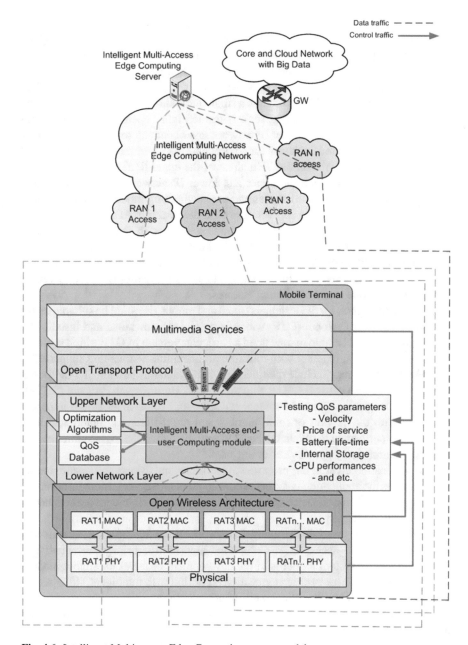

Fig. 4.6 Intelligent Multi-access Edge Computing system model

However, this chapter primarily focuses on the IP layer and above, by using various RATs below. Because the all IP framework concept is adopted in all IEEE networks, in all 3GPP 4G mobile networks, in 5G, and beyond, therefore the network layer everywhere would be IP. However, this must be separated into two sublayers: the upper and the lower IP network layer.

The upper IP network layer contains a single unified IPv6 address, which is used for routing as well as for creating sockets to the higher open transport layer and to the application layer. The lower IP network layer contains several different IPv4 (or IPv6 addresses), one IP address for a particular RAT interface. Each of these IP addresses is mapped to the unified IP address of the upper IP network layer.

The middleware between the upper and lower IP network layers contains the address of the intelligent multi-access end user computing module. This module is independent from the mobile and wireless technologies. It used to optimize packet scheduling in the 5G network (in the mobile device or in the server side) with numerous RAT interfaces or multi-access, in order to attain broadband throughput, and queue stability, simultaneously, to minimize the queue delays, etc. With one word said, the functionality of this module is to achieve superior QoS support and a high level of mobile cloud computing features.

In the upper layers, for 5G terminals (fixed and mobile) it would be suitable to use an Open Transport Protocol (OTP) which is possible to download and install. The 5G terminals would be able to download a particular version of OTP which is aimed for a particular RAT, together with multi-homing and multi-streaming support [32, 33]. The download may include modifications and adaptation version of TCP for the mobile and wireless networks, RTP, Stream Control Transmission Protocol (SCTP), Datagram Congestion Control Protocol (DCCP), and some other future transport protocol. With the support of multi-homing and multi-streaming features, a load balancing is performed to obtain the best QoS provisioning, the highest throughput, and the minimal delay per service for any user.

The application layer in our proposed 5G terminal and the general system model is the same as the application layer in the conventional OSI model. More information about the future 5G mobile terminal designs description together with the OSI layers can be found in [10, 15].

Besides, the proposed QoS routing framework in the 5G mobile terminals and the Intelligent Multi-access Edge Computing algorithm is briefly presented. The main work of the system model is in the development of novel intelligent module with advanced QoS user-centric aggregation algorithm with Intelligent Multi-access Edge Computing and with vertical multi-homing features.

This algorithm, as it was mentioned before, selects the most relevant optimization algorithm for a given heterogeneous scenario. Except the GA and LP (as in our previous works [4, 34] and [35]), for more complex problems, it can be counted on the Lyapunov optimization with the Lyapunov drift-plus-penalty framework (i.e., as in [36, 37]), which gives profound advantages regarding the key QoS parameters and better traffic control. Moreover, despite the existing frameworks and algorithm, here the system model is enriched with the possibility to combine and select different present and future optimization algorithms.

To emphasize that, the presented intelligent module simultaneously combines a few diverse traffic streams from various multimedia services that are transmitted through the same or different RAT channels. In that way it has a possibility to achieve higher throughput, higher access probability ratio, minimal end-to-end delay, stability, and optimal usage of the heterogeneous mobile and wireless resources, plus the cloud servers.

However, the Intelligent Multi-access Edge Computing processes demand higher processing power, memory, CPU, parallel processing, longer battery life, etc. In addition, according to Moore's law for the development of computers, the processing power was doubled almost every 2 years in the past decades. Therefore, the mobile terminals as computer-capable devices in terms of software and hardware follow the Moore's law.

All of this provides the possibility simultaneously to use existing and future multiple radio network interfaces for various applications. In fact, there are tunnels between the 5G mobile terminal and the Intelligent Multi-access Edge Computing server. This server may act as the main control system in the 5G core and possess the ability to collect and aggregate all traffic streams from the 5G mobile terminals, and it also performs routing which is based on given traffic policies. The Intelligent Multi-access Edge Computing server provides support of vertical multi-homing and multi-streaming, with the same protocols on open transport layer as within the 5G mobile terminal transport layer, and like that it provides many profound advantages, as well as improved traffic control.

4.5   Conclusions

The mobile and wireless technologies have resulted with several generation changes. This transformed the landscape of cellular and wireless networks into a global set of interconnected networks. The 5G technology equipment and terminals and the 5G performance levels, have already become commercially available. 5G represents a set of evolved network technologies. Its main goals are to provide unlimited mobile broadband access of data and to be able to share information at any place, at any time, and by anyone or anything for the benefit, progress and prosperity of society, businesses, and people. However, in order the systems and the equipment themselves to be labeled as 5G, it is necessary the exact levels of performance levels, regulation, and requirements to be defined.

According to presented analysis, the novel 5G MTs with Intelligent Multi-access Edge Computing and vertical multi-homing has many advantages under various network conditions. Consequently, the presented concepts and platforms for Intelligent Multi-access Edge Computing and QoS mechanisms provide the highest and the best level of access probability ratio, user throughput, the best number of satisfied consumers of mobile services, the minimum service cost, and optimized utilization of network resources as a result of sharing the traffic load. Our analysis showed that the performance gain with Intelligent Multi-access Edge Computing module in the 5G MT is higher if there are more available radio access points in comparison with the scenarios with a low number of radio access points. Also, our analysis can further be easily generalized in many multi-access heterogeneous 5G network scenarios, including any currently available or and future RATs.

Moreover, cloud computing and QoS orchestration mechanisms would be diffused in all network entities, including the smart mobile edge devices, i.e., the cloud computing would be extended at the edge of the network in a form of Intelligent Multi-access Edge Computing. A lot of functionalities in a form of virtual network would be performed in a mobile edge computing environment. This would give mobiquitous benefit to the clients. The smart devices and also smart things would be able to create and use data, services, and applications, i.e., they would be able to provide anything as a service (AaaS).

Finally, the advantages of novel QoS mechanisms and Intelligent Multi-access Edge Computing platforms, together with many other variations of them, undoubtedly are part of the nowadays and future mobile and wireless paradigm, architecture, and networks.

References

1. T. Janevski, *QoS for Fixed and Mobile Ultra-Broadband* (Wiley-IEEE Press, 2019)
2. S. Kitanov, T. Janevski, Fog Computing Service Orchestration Mechanisms for 5G Networks. J. Internet Technol. ISSN 1607-9264, Taiwan (2018)
3. J. Rodriguez, *Fundamentals of 5G Mobile Networks* (Wiley, 2015)
4. T. Shuminoski, S. Kitanov, T. Janevski (2018). Advanced QoS provisioning and mobile fog computing for 5G. Wireless Commun. Mobile Comput. J., Hindawi and Wiley
5. F. Boccardi et al., Five disruptive technology directions for 5G. IEEE Commun. Mag. **52**(2), 74–80 (2014)
6. D. Guinard, V. Trifa, et al., From the internet of things to the web of things: Resource-oriented architecture and best practices, in *Architecting the Internet of Things*, (Springer, Berlin/Heidelberg, 2011), pp. 97–129
7. B. Brech, J. Jamison, L. Shao, G. Wightwick, *The Interconnecting of Everything* (IBM Corporation, 2013)
8. N. Bhushan et al., Network densification: The dominant theme for wireless evolution into 5G. IEEE Commun. Mag. **52**(2), 82–89 (2014)
9. T. Janevski, *5G mobile phone concept. IEEE Consumer Communications and Networking Conference (CCNC) 2009*, Las Vegas, USA (2009)
10. B. Bangerter, S. Talwar, R. Arefi, K. Stewart, Networks and devices for the 5G era. IEEE Commun. Mag. **52**(2), 90–96 (2014)
11. C.-X. Wang et al., Cellular architecture and key technologies for 5G wireless communication networks. IEEE Commun Mag **52**(2), 122–130 (2014)
12. W. W. Lu, *An Open Baseband Processing Architecture for Future Mobile Terminals Design*, IEEE Wireless Communications (2008)
13. A. Tudzarov, T. Janevski, Design for 5G mobile network architecture. Int. J. Commun. Netw. Inf. Secur **3**(2), 112–123 (2011)
14. J. Noll, M.M.R. Chowdhury, *5G – Service Continuity in Heterogeneous Environments*, Wireless Personal Communications (2010)
15. M. Rahman, F. Mir, Fourth generation (4G) mobile networks – Features, technologies and issues, 6th IEE International Conference on 3G Mobile Communication Technologies (London, 2005), pp. 1–5

16. J. M. Pereira, Fourth generation: Now, it is personal. in *11th IEEE International Symposium on Personal, Indoor and Mobile Radio Communications (PIMRC)*, vol 2, pp. 1009–1016 (London, 2000)

17. J.G. Andrews et al., What will 5G be? IEEE J. Sel. Areas Commun. **32**(6), 1065–1082 (2014)

18. J. Rodriguez, *Fundamentals of 5G Mobile Networks* (Wiley, 2015).

19. Recommendation ITU-T Y.1541 (05/2002): Network performance objectives for IP-based services

20. Recommendation ITU-T Y.1542 (10/2010): Framework for achieving end-to-end IP performance objectives

21. A. Nakao, P. Du, Y. Kiriha, F. Granelli, A.A. Gebremariam, T. Taleb, M. Bagaa, End-to-end network slicing for 5G mobile networks. J. Inf. Process. **25**, 153–163 (2017)

22. S. Sharma, R. Miller, A. Francini, A cloud-native approach to 5G network slicing. IEEE Commun. Mag. **55**(8), 120–127 (2017)

23. X. Foukas, G. Patounas, A. Elmokashfi, M.K. Marina, Network slicing in 5G: Survey and challenges. IEEE Commun. Mag. **55**(5), 94–100 (2017)

24. X. Li, M. Samaka, H.A. Chan, D. Bhamare, L. Gupta, C. Guo, R. Jain, Network slicing for 5G: Challenges and opportunities. IEEE Internet Comput. **21**(5), 20–27 (2017)

25. R. Buyya, S. N. Srirama, Management and orchestration of network slices in *5G, Fog, Edge, and Clouds, a chapter in Fog and Edge Computing: Principles and Paradigms* (Wiley Telecom, Edition 1, 2019) pp. 79–10

26. Recommendation ITU-T Q.5001 (10/2018): Signalling requirements and architecture of intelligent edge computing

27. S. Kitanov, E. Monteiro, T. Janevski, 5G and the Fog – Survey of Related Technologies and Research Directions, Proceedings of the 18th Mediterranean IEEE Electrotechnical Conference MELECON 2016, Limassol, Cyprus (2016)

28. M. J. Neely, *Stochastic Network Optimization with Application to Communication and Queuing Systems,* Morgan and Claypool, USA (2010)

29. M. Malisoff, F. Mazenc, *Constructions of Strict Lyapunov Functions* (Springer, London, 2009)

30. L. Tassiulas, A. Ephremides, Stability properties of constrained queueing systems and scheduling policies for maximum throughput in multihop radio networks. IEEE Trans. Autom Control **37**(12), 1936 (1992)

31. M.J. Neely, E. Modiano, C.E. Rohrs, Dynamic power allocation and routing for time varying wireless networks. IEEE J. Sel Areas Commun. **23**(1), 89–103 (2005)

32. Recommendation ITU-T Y.2052 (02/2008): Framework of multi-homing in IPv6-based NGN

33. Recommendation ITU-T Y.2056 (08/2011): Framework of vertical multihoming in IPv6-based Next Generation Networks

34. T. Shuminoski, T. Janevski, 5G mobile terminals with advanced QoS-based user-centric aggregation (AQUA) for heterogeneous wireless and mobile networks. Wireless Netw. (2015) https://doi.org/10.1007/s11276-015-1047-4

35. T. Shuminoski, T. Janevski, Radio network aggregation for 5G Mobile terminals in heterogeneous wireless and Mobile networks. Wirel. Pers. Commun. **78**(2), 1211–1229 (2014)

36. T. Shuminoski, T. Janevski, Lyapunov optimization framework for 5G Mobile nodes with multi-homing. IEEE Commun. Lett. **20**(5), 1026–1029 (2016)

Chapter 5
An Efficient Interpolation Method Through Trends Prediction in Smart Power Grid

Weilong Ding (ID), **Zhe Wang** (ID), **Yanqing Xia, and Kui Ma**

5.1 Introduction

Electricity has become an indispensable resource for urban development. The smart power grid is based on an integrated, high-speed two-way communication network, through advanced sensing and measurement technologies, equipment technology, control methods, and decision support system technology applications [1]. The vision of the smart power grid includes advanced communication technologies to make power grid more efficient, reliable, secure, and resilient [2]. With the development and application of smart grid, business technicians can employ electricity consumption to do real-time business analytics for domain management and public services. During the development of Internet of Thing (IoT) technology, electronic devices have become edges of network as a significant role. The smart meters are edge of smart power grid. The records generated by smart meter became important support for real-time electricity estimation. With such smart meters, the business technicians can estimate electricity price and optimize power dispatch [3]. Moreover, government and power station agencies can achieve exact consumption of electricity. Due to the data transmission fluctuation or sensor abnormality, some records would be lost. The missing records will impact on subsequent analyses. So, how to improve the data quality by interpolating data for those missing records is necessary for business data analytics.

W. Ding · Z. Wang · Y. Xia
School of Information Science and Technology, North China University of Technology, Beijing, China

Key Laboratory on Integration and Analysis of Large-scale, Beijing, China
e-mail: dingweilong@ncut.edu.cn

K. Ma
Zhejiang University of Technology, Hangzhou, China

© Springer Nature Switzerland AG 2021
H. Gao, Y. Yin (eds.), *Intelligent Mobile Service Computing*, EAI/Springer
Innovations in Communication and Computing,
https://doi.org/10.1007/978-3-030-50184-6_5

However, we find two mainly technical problems against such data interpolation. First, the accuracy of attribute to be interpolated is hard to guarantee. We not only want to find out the missing of the records but also want to substitute them with sensible ones by trend prediction. The predictive precision is key factor. Second, low latency of the interpolation is required in domain. The process of records transmission is continuous and fast, and accordingly the interpolation has to be done as fast as passable to alleviate the influence of sequent steps. Therefore, a novel interpolation method is required to hold efficiency in time.

In this paper on electricity data from smart power grid, a data interpolation method for missing records is proposed by trend prediction through support vector regression. It is executed in a hierarchical edge environment in smart power grid. This paper is organized as follows. In the second section, the practical background and related works are introduced. In the third section, the method is elaborated. Section 5.4 shows data set and experimental results, and then a conclusion is summarized in the fifth section.

5.2 Background

5.2.1 Motivation

With the application of smart devices, lots of data have been generated. In smart grid, record transmission among smart meters is no longer point to point but forms tree-like hierarchical levels. According to the business status in Beijing, such data is collected by three levels before it was gathered in central cloud for analysis, as showed in Fig. 5.1. The records are generated by smart meters of the smart power grid. Then, the records are transmitted to community power station through the network. After that, it will be summarized by the regional power supply station. Finally, summarized data is transmitted to the municipal power supply station.

The records will be uploaded every 7 minutes in each station. The records of electricity data have the data structure as in Table 5.1. Each record has a timestamp, a numeric value of power consumption, and identification of geographical location.

Those servers compose a typical tree-like topology: one station could gather data from multiple lower servers and transfer its results to a certain upper one. This hierarchical structure increases the odds of missing records. The record missing defined as follows.

Definition 1 (Record of Electricity Data and Record Missing) In an interval of 7 minutes, D_t^l is a record generated from station l at time t. As time series, such records would be transmitted to a certain upper level station I's. The record missing in this paper implies a specific record D_i^l is lost or its attribute *consumption* is missing.

The record missing phenomenon will impact analytical results. Therefore, in order to provide high-quality data for later operation, we need to interpolate missing ones during the process of data transmission. We decide to interpolate the missing

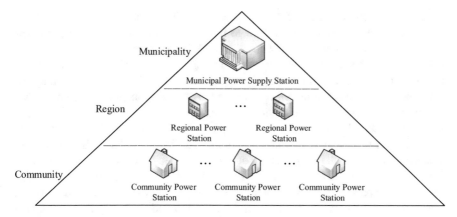

Fig. 5.1 Hierarchical structure

Table 5.1 Data structure of electricity record

Attribute	Notation	Type
ID	Record id	Number
timestamp	Generated time	Time
consumption	Electricity consumption	Aggregation
location	Electricity meter location	Space

The ID represents a unique identifier for each record. The timestamp presents the time when this record was generated. The consumption is an accumulative power consumption of a smart meter at this time, which is a monotonically increased numerical value. The location implies the spatial identifier of the current smart meter.

records through trends prediction and call it data interpolation method. Because of the high demand on time sensitiveness, the data interpolation ought to be completed fast enough. To improve processing latency, Nazmudeen et al. [4] argued the fog computing moving calculation down to the edges of the network. This is similar to the basic idea in this paper. We can apply the data interpolation method in each layer of smart power grid. That will disperse the stress of low latency from center to edge.

5.2.2 Related Work

Record missing is common phenomena in data transmission. We propose a data interpolation method to calibrate the missing record. So, a suitable data clean procedure is required to find out the record missing, and then the predicted result is used as the missing ones. We need to know how data cleaning procedure works and find a suitable method to predict reasonable and accurate results with low latency. Related work can be divided into two perspectives.

One perspective is data clean pre-processing. This step is to find out record missing. There are researches on how to find and drop the illegal records but few of them on how to interpolate them. Xia et al. [5] show how to handle highway domain data. They found abnormal data during the data clean procedure and use historical data to fix it. Ding et al. [6] propose a model to predict the missing travel-time data in stream environment and reduce latency. Smart meter, as the end device of smart grid, generated the electricity records continuously. The amount of data is gradually accumulated, and the data interpolation has to be low intrusive for data transmission. Low latency interpolation in tree-like hierarchy has to be considered. The work of Bose et al. [7] uploads all of our grid data to the central cloud. In today's big data condition, the amount of data has exploded. Edge computing is a new type of computing paradigm. It would be more efficient to process the data at the edges of the network [8]. There are other works for edge computing [9, 10]. Ding et al. [11] use edge computing in his research. The data center and sensor constitute cloud-edge environment. They propose data clean model on each edge of the network. However, there are not many studies in the domain of smart grids. The power grid structure is also edge environment. We introduce our work on a hierarchical computing architecture so that the data can be computed on the production side to speed up the efficiency of data processing.

The other perspective is data interpolation through trends prediction. The most important part here is prediction. In recent years, many scholars have proposed prediction models in power grid. Qiu-Hua et al. [12] proposed gray model to forecast electricity sales. The gray model theory is a mathematical model based on associated space and smooth discrete function. It has more requirements on the smoothness of the data and weaker generalization ability. Our electricity data is increment by itself, but the smoothness cannot be guaranteed. Wei et al. [13] proposed a power consumption prediction using a time series differential autoregressive moving average model (ARIMA). The ARIMA model is simple, small on samples, and has high prediction accuracy, but the execution time is longer. In our scenario, the low latency is also required imperatively. Since the rise of machine learning, many studies emerge. Sozen et al. [14] use artificial neural networks (ANN) to predict Turkey's net energy consumption in 2005. Hui et al. [15] use back propagation artificial neural network to predict the consumption of Changchun City. Dawen et al. [16] proposed XGBoost algorithm to predict power consumption, and the effect is also obvious. They trained the model through the consumption data in recent years and finally got better results. However, compared to their practical background, the problem we face is insufficient training data for the model. We don't have large size data set for training. Not every layer in the grid has the capability to store quantity data. Support vector regression (SVR) on power consumption data [17–21] is proposed either. They all achieved good accuracy, especially on small samples. Such methods on support vector machines (SVM) are learning tools based on statistical learning theory (SLT) and structural risk minimization (SRM) [18]. Support vector regression (SVR) is one category of SVM, which supports small samples well and can quickly construct the model. It fits with our motivation and purpose. Therefore, we will employ the SVR to predict the short-term trends of electricity consumption for data interpolation.

5.3   Methods and Models

5.3.1   *Methodology*

The work is originated from smart power grid of Beijing, China. The record generated by smart meter will be transmitted to community power station. After aggregative operations, the records would be sent to power station of upper level, as shown in Fig. 5.1. In this paper, the proposed interpolation method for missing records of smart meter data is defined as follows.

Definition 2 (Data Interpolation) For a missing record D_t^l, data interpolation is the procedure to interpolate a record $\hat{D}_t^l$ as substitute. The interpolated record contains predicted value of attribute *consumption* at time t from station l got through trends prediction model.

The overview of our method is shown in Fig. 5.2 below.

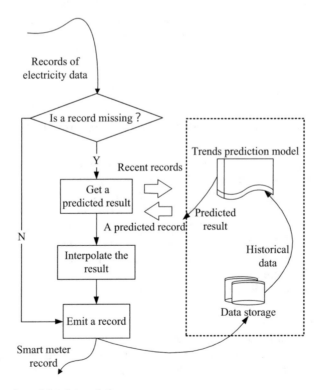

Fig. 5.2 Overview of data interpolation

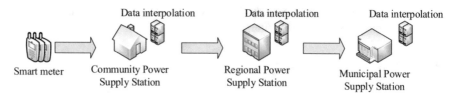

Fig. 5.3 Data interpolation method deployed in hierarchical edge environment

Due to constant data generation, the record missing can be found by timestamps of sequent records. If one record D_i^l is missing, we can calibrate *location* and *timestamp* attribute due to temporal correlation. But for the attribute *consumption*, we can't calibrate it directly. Base on SVR, the trends can be predicted for the missing attribute. In our trend prediction method, the first step is data pre-processing. Because the *consumption* attribute is an accumulative value, so the difference between the attributes in two successive records used for predicted would be more efficient than that of the raw one. After normalization on the range of different attributes, prediction is executed on the constructed feature vectors. After training and testing, the predicted result can be found to substitute the missing D_i^l.

After completing the data interpolation, we also need to find a suitable way to deploy it. Due to the special hierarchy, the data flow can be regarded as the data transmission in Fig. 5.1, and stations in each layer can be regarded as edge servers. Our data interpolation method is deployed in each layer to reduce the analytical latency. Figure 5.3 shows the deployed environment in hierarchical edge for our method.

5.3.2   Trend Prediction to Substitute Missing Records

For data interpolation, we use trend-predicted results for missing records' substitution. Smart meters generate data continuously with strong spatiotemporal correlation, and deep learning methods are not suitable in real-time data condition. SVM has good support for small data set and response fast. We can use small training set to build the model, and the time for training and prediction is competitive. So, we select support vector as the theoretical basis and use support vector regression for trend prediction. In this way, historical data within a certain time range can be used for model training. When the model has been trained, the predicted results would be interpolated by writing back the corresponding encapsulated record into data flow.

The SVM was largely developed at AT&T Bell Laboratories by Vapnik and co-workers [22]. It is based on the principle of structural risk minimization, which better fits the "over-learning" and has good generalization ability [1]. The theory on support vector evolves two different directions as classification and regression with respective models. SVR used for the regression on continuous values is adopted here for trends prediction.

For the data interpolation, how to construct feature vector is the most important phase next. The feature vector is constructed in a certain period whose length is fixed during the data transmission. The definition of feature vector is shown as follows:

Definition 3 (Feature Vector of Electricity Consumption) With given time range $[t - T - 1 .. t - 1]$, the feature vector of attribute *consumption* in missing record D_i^l can be described as $C_t^l = (C_{t-1}^l, C_{t-2}^l, C_{t-3}^l, \cdots, C_{t-T-1}^l)$. Here, $C_i^l = D_i^l.consumption - D_{i-1}^l.consumption$ is an increment of that attribute. It is constructed with the time series in recent T periods.

So, for a time range T, we can get a set of feature vectors and a target value. We use this training set to training our prediction model. Our goal is to find a function $f(x)$ that can obtain targets y_i with most ε deviation for all the training inputs as flat as possible. So, we can give the description of the $f(x)$ in the form as

$$f(x) = \langle w, x \rangle + b, w \in \aleph, b \in \Re \tag{5.1}$$

Here, $\langle ., . \rangle$ denote the dot product in $\aleph$. Then, as we know from (5.1), w is the weight matrix as the coefficient of the input vector x. Model parameters w and b will be automatically determined after training. For the optimal function $f(x)$, we need to minimize the $\|w\|$, as well as $\|w\|^2$. In order to tolerate errors in our data set, slack variables ξ_i, ξ_i^* and constant C are introduced. The optimization procedure can be described as follows:

$$\text{minimize } \frac{1}{2}\|w\|^2 + C \sum_{i=1}^{n} (\xi_i + \xi_i^*) \text{ subject to } \begin{cases} y_i - \langle w, x_i \rangle - b \le \varepsilon + \xi_i \\ \langle w, x_i \rangle + b - y_i \le \varepsilon + \xi_i^* \\ \xi_i, \xi_i^* \ge 0 \end{cases}$$
$$\tag{5.2}$$

So that the loss function of model can described by follows:

$$|\xi|_\varepsilon = \begin{cases} 0 & if \ |\xi| < \varepsilon \\ |\xi| - \varepsilon & otherwise \end{cases} \tag{5.3}$$

After all these preparations, we can use convex quadratic programming to find the support vector and form an optimal hyperplane. That is the principle of our trends prediction.

5.3.3   Trend Prediction Implementation

In data interpolation by trend prediction, penalty coefficient C, kernel function k, and loss parameter ε are the hyper-parameters we want to tune. Penalty coefficient C in

(5.2) is the balance factor between the flatness $f(x)$ and deviations larger than ε. We set a constant C for the regression equation. The constant C represents the tolerance of the model to the bias data. When C is larger, the lower the tolerance, the more support vectors will be, and the regression will become more complex with less stability. Conversely, the regression equation will become smoother with stability. So, we use cross validation to find the most appropriate C and ε. In order to get better result, we use kernel function k to map features to high dimensions. After multiple comparisons, we use radial basis function (RBF) for high-dimensional mapping, which will result a better effect. RBF kernel function is the most commonly used support vector mapping function, which can be expressed as $k(x_0, x_1) = \exp\left(\frac{d(x_0, x_1)^2}{2*\sigma^2}\right)$. The hyper-parameter we focus is gamma, described as gamma $= \frac{1}{2*\sigma^2}$. The hyper-parameter gamma implicitly determines the distribution of data after mapping to a new feature space. The larger the gamma is, the less the support vectors are required; the smaller the gamma value is, the more support vectors are required. The number of support vectors affects the speed of training and prediction. In this experiment, we set the value of gamma to $\frac{1}{T*T.\mathrm{var}()}$.

So, the implementation of our data interpolation method is shown in Table 5.2.

First, we use cross validation to determine the hyper-parameters. On constructed feature vectors, the prediction model is trained. Next, we utilize Lagrange multipliers to convert to standard dualization method to find the function. After completing the training, we get a regression function $f(x)$, which is the final hyperplane we obtain through the support vector regression. This regression function can well describe the correspondence between electricity consumption and spatiotemporal factors. Next, we use test set to get the predicted increment value, and we sum with

Table 5.2 Data interpolation method implementation

Algorithm 1 *data interpolation method based through SVR*
Input: a record R, given time range T, historical data D in time range $2T$, penalty coefficient C, kernel function k, loss parameter ε Output: a substitution record R_c
1. Cross validation to determine C, ε
2. If R is judged as missing according to Definition 1:
3.　　Read-in D, calculate electric consumption increment $C^l_{in_i}$ on D through time range T and do normalization
4.　　Construct feature vectors as x and training label $f(x)$
5.　　Train SVR model on those vectors
6.　　Through training model, predict the increment, sum with the latest data, and form current consumption data
7.　　Interpolation that data into record R as *consumption* attribute
8.　　If *location* and *timestamp* of R is missing, interpolate its base temporal correlation, form R_c
9.　　Interpolate R_c to data flow
10. End if

latest consumption attribute to form a substituted record. Finally, we interpolate the missing record and write back to data flow when record missing found.

5.4   Experiments

5.4.1   Environment

In order to evaluate data interpolation method that we proposed, we deploy the its implementation in a hieratical edge environment as Sect. 5.2.1. Three servers are used to simulate a tree-like hierarchical edge environment. Each of them has eight processors, 16G memory, and 32T storage. For the regional layer, one server is utilized to receive community layer data. For the community layer, two independent servers are configured to send aggregated data to regional layer and receive records from smart meter. Our data interpolation method works as edge service of the hieratical structure. For each server, our dedicated data generator [23, 24] is used to replay and simulate stream data. The data we imported was generated by real smart meter in Beijing since Jun 1, 2018 to Jun 30, 2018. The concurrency and data rate of the simulated data can be configured on demand by our data generator. By default, the concurrency is about 20, which depends on lower layer station number. The data rate is one record per 7 minutes, as the same as practical scene. All experiments are performed on the community layer.

In any edge server, some tools are deployed. Scikit-learn library with Python-3.6.6 is installed to support scientific computation and machine-learning algorithms. Dependent library like NumPy, pandas are also installed.

5.4.2   Evaluation

Two experiments are designed here. On the one hand, to evaluate prediction effects, we adopt absolute percent error (APE), maximum error (MaxError), mean absolute percent error (MAPE), mean absolute error (MAE), mean square error (MSE), root mean square error (RMSE), and R-squared for evaluation. In data interpolation method, the time range T is set as 10, the slack variable ε is taken as 0.089, kernel function is RBF, and the penalty coefficient C is 10. All training samples satisfy the KKT condition as the algorithm stop condition. On the other hand, to evaluate performance cost, we change the data rate parameter of data generator to compare the difference of executive time between original process and our data interpolation method.

Experiment 1: Prediction Effects   The data generated in Beijing on Jun 11, 2018 is replayed by data generator. Besides our interpolation method (abbr. SVR) is deployed in the service of any edge server, linear model ARIMA [13] is also implemented and deployed for comparison. The executive latencies for a trend prediction are counted during time elapses. All the metrics between ground truth values and predictive ones are computed after finishing data replay.

The predictive effects are illustrated as Fig. 5.4 and Table 5.3. Figure 5.4a shows that the predicted value obtained by our prediction and counterpart has similar values. From Fig. 5.4b, we can see APE of both SVR and ARIMA is under 0.005. The performance criterion in Table 5.3 tells us two prediction methods can hold high accuracy as expectation.

But from the perspective of executive time of trends prediction in Fig. 5.5, two prediction models present distinctly. Although execution times of these two are under 12ms, SVR has much lower latency with stability than ARIMA. Therefore, from the experimental results, our data interpolation model base on SVR can meet the requirements in high predictive accuracy and low latency.

Experiment 2: Performance Cost In this part, by increasing data rate, the performance cost of our method is evaluated. In the same power station, we compare the time consumption of data transmission with our method (abbr. with interpolation) or without it (abbr. origin) in an edge server. We increase the data rate from 1 thousand records per second to 1000 thousands records per second. After the data generating smoothly, we record the average execution time under each data rate in such two configurations.

The results can be seen from Fig. 5.6, and the time will grow when data rate increases. With the data interpolation method, the extra required time than original one is less than 1 second. That is acceptable compared with the time quantity of the original one. Therefore, with the increase of data rate, no clear burden exists in performance of time and proves the reliability of our method.

5.5   Conclusion

With the prosperousness of smart grid, business analyses require accurate data generated by smart meters. But records missing will hinder further exact analytical explanation. In order to improve the quality of the data against record missing, interpolation data method faces challenges in accuracy and latency guarantee. We propose a data interpolation method through trend prediction by SVR model and apply it in layers of the smart power grid to ensure low latency. Experiments show that the accuracy of predicted results is high enough compared with the real data in second level executive time. It is verified that the extra time cost by our method is limited than the original one.

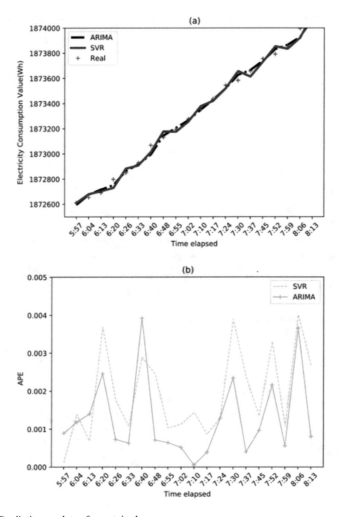

Fig. 5.4 Prediction on data of a certain day

Table 5.3 Predictive error for comparison

Performance criterion	MaxError	MAPE	MAE	MSE	RMSE	R-squared (%)
Support vector regression	75.13	0.0019	35.975	1738.4	41.69	99.14
ARIMA	73.44	0.0013	24.055	966.07	31.08	99.52

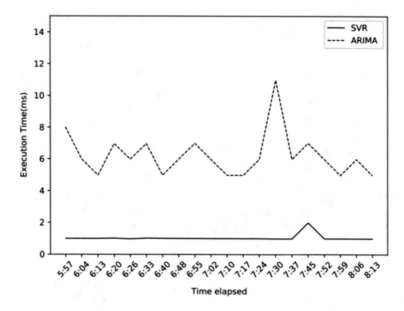

Fig. 5.5 Executive latency with elapsed times

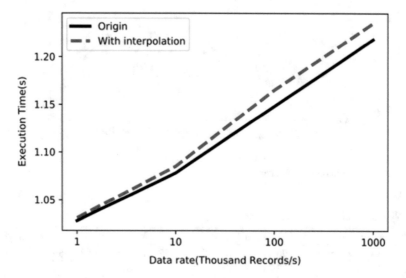

Fig. 5.6 Executive latency with data rate

Acknowledgments This work was supported by National Natural Science Foundation of China (No. 61702014), Beijing Municipal Natural Science Foundation (No. 4192020), and the Top Young Innovative Talents of North China University of Technology (No. XN018022).

References

1. S. Jianxin, Y. Shaniin, Electricity consumption prediction based on SVR with particle swarm optimization in smart grid [J]. Sci. Technol. Manage. Res (2013)
2. J. Lin, W. Yu, X. Yang, Towards multistep electricity prices in smart grid electricity markets [J]. IEEE Trans. Parallel Distrib. Syst. **27**(1), 286–302 (2015)
3. J. Lin, W. Yu, X. Yang, Towards multistep electricity prices in smart grid electricity markets [J]. IEEE Trans. Parallel Distrib. Syst. **27**(1), 286–302 (2015)
4. M.S.H. Nazmudeen, A.T. Wan, S.M. Buhari, *Improved Throughput for Power Line Communication (PLC) for Smart Meters Using Fog Computing Based Data Aggregation Approach [C].* Smart Cities Conference (IEEE, 2016)
5. Y. Xia, X. Wang, W. Ding, A data cleaning service on massive spatio-temporal data in highway domain, in *Service-Oriented Computing – ICSOC 2018 Workshops*, ICSOC 2018. Lecture Notes in Computer Science, ed. by X. Liu et al., vol. 11434, (Springer, Cham, 2019)
6. W. Ding, S. Zhang, Z. Zhao, A collaborative calculation on real-time stream in smart cities [J]. Simul. Modell. Pract. Theory **73**, 72–82 (2017)
7. A. Bose, Smart transmission grid applications and their supporting infrastructure [J]. IEEE Trans. Smart Grid **1**(1), 11–19 (2010)
8. W. Shi, I.E.E.E. Fellow, et al., Edge computing: Vision and challenges [J]. IEEE Internet of Things J. **3**(5), 637–646 (2016)
9. W. Dewen, Y.A.N.G. Liping, Stream processing method and condition monitoring anomaly detection for big data in smart grid [J]. Autom. Electr. Power Syst. **40**(14), 122 (2016)
10. S. Weisong, S. Hui, C. Jie, et al., Edge computing—An emerging computing model for the internet of everything era [J]. J. Comput. Res. Dev **54**, 907–924 (2017)
11. W. Ding, Z. Zhao, DS-harmonizer: A harmonization service on spatiotemporal data stream in edge computing environment [J]. Wirel. Commun. Mobile Comput. **2018**, 1–12 (2018)
12. L. Qiu-Hua, C. Jie, G. Hai-Qing, Forecasting on the amount of sales of electric power based on improved grey model [J]. Stat. Inform. Forum (2009)
13. Y. Wei, C. Chao, X. Bin, et al., Forecasting for monthly electricity consumption using X12 multiplication method and ARIMA model [J]. Proc CSU-EPSA (2016)
14. A. Sozen, E. Arcaklioglu, M. Ozkaymak, Modelling of the Turkey's net energy consumption using artificial neural network. Int. J. Comput. Appl. Technol. **22**(2/3) (2005)
15. Z. Hui, W. Wei, L.I. Xiao-Mei, Study of the model for forecasting vendition kW·h Using ANN [J]. Hunan Electr. Power (2004)
16. H. Dawen, F. Penglan, Power consumption forecasting application based on XGBoost algorithm [J]. Modern Inform. Technol. (2017)
17. K. Kavaklioglu, Modeling and prediction of Turkey's electricity consumption using support vector regression [J]. Appl. Energy **88**(1), 368–375 (2011)
18. G. Yan-Dong, L.I. Rong, Annual electric demand forecasting based on support vector regression [J]. Techniq. Autom. Appl (2008)
19. L.L. Li, Q.Z. Jing, C.H. Xin, et al., Forecast of electric power demand in Hubei province based on SVR model during the period of 2016—2010 [J]. Power Demand Side Management (2017)
20. F. Kaytez, M.C. Taplamacioglu, E. Cam, et al., Forecasting electricity consumption: A comparison of regression analysis, neural networks and least squares support vector machines [J]. Int. J. Electr. Power Energy Syst **67**(67), 431–438 (2015)

21. M.S. Al-Musaylh, R.C. Deo, J.F. Adamowski, Y. Li, Short-term electricity demand forecasting with MARS, SVR and ARIMA models using aggregated demand data in Queensland, Australia [J]. Adv. Eng. Inform. **35**, 1):1–1)16 (2018)
22. A.J. Smola, B. Schölkopf, A tutorial on support vector regression. Stat. Comput. **14**, 199–222 (2004)
23. W. Ding, Y. Han, J. Wang, Z. Zhao, Feature-based high availability mechanism for quantile tasks in real-time data stream processing. Softw. Pract. Exp **44**(7), 855–871 (2014). 26
24. W. Ding, Z. Zhao, and Y. Han, A Framework to Improve the availability of Stream Computing. in *Proceedings of the 2016 23rd IEEE International Conference on Web Services* (ICWS 2016), pp. 594–601, IEEE, SanFrancisco, June 2016

Chapter 6
2PC+: A High Performance Protocol for Distributed Transactions of Micro-service Architecture

Pan Fan, Jing Liu, Wei Yin, Hui Wang, Xiaohong Chen, and Haiying Sun

6.1 Introduction

With the rapid development of the Internet field, the traditional single-service architecture is facing huge challenges. Micro-service architecture [1] has become extremely popular today. The core of the micro-service is to implement separate deployment, operation and maintenance, and expansion according to the business module. Unfortunately, there are many new problems in micro-services compared to traditional single-service architectures. Among them, distributed transaction in micro-services is one of the most common challenge [2].

In the traditional single-service architecture, whole the business modules are concentrated on the same data source. Therefore, it is convenient to implement the local transaction mechanism to ensure data consistency within the system, such as 2PC [3], MVCC, etc. However, in the micro-service architecture, service modules often straddle heterogeneous distributed system, and they are deployed across services and resources [1]. With the rapid growth of the number of micro-service nodes, 2PC and MVCC exist in their performance bottlenecks. So that 2PC and MVCC are hard to reach the processing performance of micro-services. In the evaluation experiment in Section 4, the latency of 2PC dropped below 35% of its maximum as contention increased.

P. Fan · J. Liu (✉) · W. Yin · X. Chen · H. Sun
East China Normal University, Shanghai, China
e-mail: jliu@sei.ecnu.edu.cn; yin_wei@careri.com; xhchen@sei.ecnu.edu.cn; hysun@sei.ecnu.edu.cn

H. Wang
Shanghai Avionics co. Ltd., Shanghai, China
e-mail: wang_hui@careri.com

In this paper, a distributed transaction concurrency control optimization protocol 2PC+ is proposed, which can extract more concurrency in the case of high competition than previous methods. 2PC+ is based on the traditional two-phase commit protocol, combined with transaction thread synchronization blocking optimization algorithm SAOLA.

We apply proposed 2PC+ protocol to the case of MSECP micro-service platform. Through experimental result, 2PC+ has obvious improvement in RT and TPS performance compared to 2PC for distributed transaction.

The rest of this paper is organized as follows. Section 2 elaborates the basic concepts and knowledge background in distributed transactions. In Section 3, we introduce the design details of the 2PC+ protocol and an optimization algorithms SAOLA based on 2PC protocol. The TLA+ verification process of the SAOLA is given in Section 4, and its feasibility is verified by running the results of TLC. Section 5 gives a comparison experiment between 2PC+ and 2PC. Section 6 discusses some related work, and Section 7 summarizes this paper.

6.2   Preliminaries

2PC and OCC Low-Performance Approaches Developers prefer the strongest isolation level serializability, in order to simplify the process of controlling distributed transactions. To guarantee the rules in transaction, traditional distributed systems (e.g. relational database) typically run standard concurrency control schemes, such as two-phase commit (2PC) combined with optimistic concurrency control (OCC) [11]. However, in many conflict transaction scenarios, 2PC combined with OCC measures perform poorly under highly competitive workloads.

For example, Table 6.1 shows a snippet of a new order transaction that simulates a customer purchasing two items from the same store. The transaction consists of two threads working on fragments P1 and P2, each of which reduces the inventory of different materials. Although each fragment can be executed automatically on its own machine, distributed control is still required to prevent fragmentation of the fragments between services. For example, suppose we keep the inventory of goods a and b constants and always sell the two together. In the absence of distributed transaction control, one customer can buy a but not b, while another customer can buy b but not a. This is due to the presence of locks in 2PC, so transactions can be aborted or even fail due to long thread blocks.

X/Open DTP X/Open [4] is the most widely used distributed transaction solution model in the single architecture. The key point is to provide a distributed transaction specification protocol: XA protocol. It uses the 2PC protocol to manage distributed transactions. The XA protocol specifies a set of communication interfaces between the resource manager (RM) and the transaction manager (TM). It is mainly composed of three main modules, i.e., RM, TM, and application (AP). Among them, RM is specifically responsible for the resource groups actually involved in

Table 6.1 A fragment of new-order transaction containing two pieces

transaction new_order_fragment
#simplified new-order "buys" 1 of a, b
input: a and b
begin
...
P_1:
R(tab=" Inventory ", key= a) $\rightarrow$ number
if (number > 1):
W(name=" Inventory ", key= a) $\leftarrow$ number - 1 ...
P_2:
R(name=" Inventory ", key= b) $\rightarrow$ number
if (number > 1):
W(name=" Inventory ", key= b) $\leftarrow$ number - 1 ...
end
...

the system, such as database resources and system operation platform resources. TM controls the distributed transaction process globally within the system, including the execution cycle of distributed transactions and coordination of RM resources.

6.3   Design

Based on the defects in the distributed transaction solution under the 2PC protocol, and optimizing the processing performance in the micro-service architecture for its problem, we propose an optimized 2PC+ protocol. It highly reduces the time cost of thread synchronization blocking when the service node processes distributed transaction in the system.

In the process based on the 2PC protocol, the thread participating in the transaction needs to be blocked after the two commits. When all the participants have completed the commit transaction, the blocking lock can be released. Thread synchronization blocking optimization in resource manager is the key to improving the performance of the original solution. In this section, the algorithm SAOLA is proposed and given the specific design and implementation.

Percolator Transaction Percolator is developed by Google to handle incremental web indexing and provides a strong and consistent way to update indexing information in a cluster of machines under distributed systems [5]. Two basic services are provided in percolator transaction, namely, a timestamp identifying the order of transactions and a distributed lock that detects the state of the process. It applies a single-row atomicity mechanism in big table. The commit and rollback operations in the original multi-row, multi-column distributed transaction are converted to simple single-line transactions [6]. Percolator uses lock, data, and write to execute the process, lock to store data information, write to save the final write data, and data to save the current timestamp version.

The Two-Level Asynchronous Lock Mechanism Converting synchronous blocking in a transaction commit process to asynchronous non-blocking is key in the optimization process. We refined the fine-grained of the lock into a second-level optimistic lock [7, 8] and proposed this optimization algorithm SAOLA (Secondary Asynchronous Optimistic-Lock Algorithm). The specific implementation of the SAOLA algorithm is shown below:

1. Initialize settings in the transaction properties. It includes *value*, which is the current actual value of the transaction; *beginVersion*, which is the version serial number of the transaction at beginning; *commitVersion*, which is the version serial number of the transaction commit; and *lock*, which stores the uncommitted transaction, and lock is divided into two different level locks, namely, *firstLock* and *secondLock*. They represent the two phases of the lock, respectively. And specify that the *secondLock* includes the *firstLock* information.
2. BeginTransaction. As shown in Table 6.2. At the beginning, transaction object T_1 obtains the value of *beginVersion* as b_v and then determines whether there is a lock in the transaction object. If it does not exist, try to obtain some latest version directly from b_v. And the committed transaction obtains its current latest data value through its *beginVersion*. Otherwise, if there is a lock, the following three cases will occur:

   - There is another transaction object T_2 is committing, the data value of T_2 is locked, and then it is necessary to wait for T_2 to complete the commit, finally polling for retry (the number of polling can be configured).
   - In the first case, if the T_1 wait time has passed a certain threshold WAIT_TIME, the value of T_2 is still locked. It can be determined that T_2 has network fluctuations or unforeseen exceptions such as service downtime, and it is directly considered that T_2 has been interrupted. In this condition, T_1 can release the lock.
   - The *lock* in T_2 may be abnormally cleared without exception. T_2's *firstLock* has completed committing and has been successfully released. However, an exception occurred in *secondLock* that caused unsuccessful commit and remains. At this condition, it can release the *lock* directly.

3. PreCommit. The algorithm is shown in Table 6.3. At this time, T_1 has obtained all the latest values and can start executing transaction precommit. The process is divided into the following three branches:

   - The first case is for all transaction objects (only T_1 and T_2 transaction objects are assumed). The *value* of T_1 and T_2 is after the version sequence b_v, and it is judged whether there is a write operation of other transaction objects T_x. If it exists, then T_x has updated the latest data value, and the current T_1 and T_2 are directly rolled back, so the process can be ended.

Table 6.2 Algorithm of BeginTransaction

Algorithm 1 BeginTransaction

Input List<TransactionItem> txGroup

Output：newestValue

1: **for** T_{start} *and* T_{else} **to** txGroup

2: **if** T_{start}.isLocked == TRUE **then**

3:    *waiting for* T_{else} *update to commit*

4:    **if** *poll until* T_{start}.isLocked == FALSE *within* WAIT_TIME

5:      break;

6:    **end if**

7:   **else**

8:    *releaseLock*(T_{start})

9: **if** !firstLock. isLocked && secondLock.isLocked **then**

10:    *releaseLock*(T_{start})

11: **end if**

12: **end if**

13: **else** // 不存在lock

14: **Long** b_v = T_{start}.beginVersion

15: newestValue = *getByBeginVersion*(b_v)

16: **return** newestValue

Table 6.3 Algorithm implementation of PreCommit

Algorithm 2 PreCommit

 # items represents the list of all transaction objects

1: **for** T **to** items

2: **If** T.hasWriteData() == true || T.isLocked() == true **then**

3:   group.doRollBack()

4:   **return**

5: **end if**

6: **else**

7:   T.isLocked = true

8:   *write to* newestItem.value to T

9:   group.doCommit ()

10: **end else**

Table 6.4 Algorithm implementation of Second-Commit

Algorithm 3 Second-Commit
items represents the list of all transaction objects
1: **for** T **to** items
2:   **if** T.firstLock.isLocked() == true **then**
3:     group.commit ()
4:   **end if**
5:   **else**
6:     group.rollBack()
7:     T.secondLock.ansyReleaseLock()
8: **end else**

- The second case is to judge whether the locks in T_1 and T_2 are locked. If they match, both T_1 and T_2 need to be rolled back.
- The third case is that there is no write of the new object T_x; T_1 and T_2 do not exist *lock*. The *firstLock* of them is set to locked, and the latest data *value* is written, so the transaction group can be committed.

4. Second-Commit. The algorithm is shown in Table 6.4. After completing the precommit, all transaction objects (assuming that there are still only T_1 and T_2) can perform the final two-phase commit step. First, T_1 and T_2 will judge whether their *firstLock* is in the locked state. If they match, they can commit the transaction. Otherwise, it represents that another has cleared the *firstLock* of T_1 or T_2 and then execute the rollback operation.

   The transaction's main process has been completed. For *secondLock*, it can be completely separated from the main process, using thread asynchronous mode to clear *secondLock* and commit the transaction. Thus, even if an exception occurs in the operation step, T_{next} for the next transaction object, it finds that *firstLock* in T_1 or T_2 has been cleared, but *secondLock* still exists; T_{next} will automatically clear *secondLock* for T_1 or T_2 and commit the transaction.

6.4 Correctness

The formalized method TLA+ language is applied for verify the optimized scheme, and the rigorous mathematical logic is used to detect the logical feasibility of the optimized algorithm SAOLA. We give the TLA+ verification steps as follows:

1. Set the two invariants in the distributed transaction to constant: the current actual value of all participating transaction and all participant transaction RMs. TLA+ can be expressed as:

   CONSTANTS   VALUE   CONSTANTS   RM

2. Set variables in the process: transaction status of the occurrence of RM, represented by the variable *rm_status*, i.e., "beginning," "preparing," "precommit," "committed," "cancel," etc.; The variable *rm_v* represents RM's current version and the global version sequence *ascend_v*. For the two locks of the control version in RM, i.e. *firstLock* and *secondLock*, represented by the variable rm_lock, two versions of the sequence appearing successively in the transaction flow *beginVersion* and *commitVersion* are represented by the variables *begin_v* and *commit_v*; respectively, the variables *first_val* and *second_val* are used to represent the RM values corresponding to the two versions; the variable *rm_data* is used to save the current actual value in the RM. Finally, the *committed_v* is used to save the committed transaction version.

```
VARIABLES    rm_status       VARIABLES    rm_v
VARIABLES    rm_lock         VARIABLES    begin_v
VARIABLES    commit_v        VARIABLES    rm_value
VARIABLES    ascend_v        VARIABLES    committed_v
```

3. Next, initializing the values of the individual variables. At the beginning, all variables are staying at initial value.

```
Init ==
    /\   rm_status = [r \in RM|-> "beginning"]
    /\   rm_v = [r \in RM| -> [begin_v |-> 0, commit_v |-> 0]]
    /\   rm_value = [r \in RM| -> [first_val|-> "",]]
    /\   second_val|->""]]
    /\   rm_lock = [v \in VALUE |->{}]
    /\   rm_data = [v \in VALUE |-> {}]
    /\   ascend_v = 0
    /\   committed_value = [v \in VALUE |-> <<>>]
```

4. When the process begins, the variable transaction state *rm_status* is "beginning," and the next state of *rm_status* does not contain "preparing." The *ascend_v* is in an ascending state, as shown in line 3. And the next state of the constraint *rm_value* cannot satisfy the getVal condition. In the next state of the current version of the RM, the start *versionbegin_v* is not equal to the next state of the version, as shown in the last line in TLA+.

```
Begin(r) ==
    /\   rm_status[r] = "beginning"
    /\   rm_status' = [rm_status   EXCEPT ![r] = "preparing "]
    /\   ascend_v' = ascend_v + 1
    /\   rm_value' =   [rm_value   EXCEPT ![r] = getVal]
    /\   rm_v' = [rm_v   EXCEPT !. begin_v = ascend_v']
```

5. The TLA+ below indicates that the transaction initially loads the process. At this time, *rm_status* stays at the "preparing" state. If the *preCommit* can be executed in the stage, and the next state of the *rm_status* cannot be "precommit"; otherwise, it is judged whether the resettable lock *isResetLock* condition is satisfied: if so, the reset lock can be executed.

```
Loading(r) ==
    /\   rm_status[r] = "preparing"
```

```
/\  IF   isPreCommit(r)        THEN
          /\   rm_status ' = [rm_status EXCEPT ![r] "]
          /\   = "precommit"]
         ELSE   IF   isResetLock (r)        THEN
          /\   resetLock(r)
         ELSE
              /\   rm_status' = [rm_status EXCEPT ![r]
              /\   = "cancel"]
```

6. TLA+ statement of the transaction precommit process. At this time, *rm_status* is the "precommit" state. If the final commit process *canCommit* can be executed, the global version *ascend_v* is incremented. The next state of *rm_v* is not allowed to be *ascend_v*, and the next state of the constraint *rm_status* does not contain "committing." Then it determines whether the RM can lock all values, i.e., *isAllLock*: if it matches, lock it. Otherwise, the constraint *rm_status* next step state does not contain "cancel" state. The TLA+ language in Commit is shown below, representing the two-phase commit step of the algorithm.

```
PreCommit(r)  ==
      /\   rm_status[r] = " precommit "
      /\   IF   canCommit(r)        THEN
          /\   ascend_v' = ascend_v + 1
          /\   rm_v' = [rm_v  EXCEPT !. commit_v
          = ascend_v']
          /\   rm_status ' = [rm_status  EXCEPT ![r]
                    "committing"]
          ELSE   IF   isAllLock(r)        THEN
          /\   allLock(r)
      ELSE
          /\   rm_status ' = [rm_status EXCEPT ![r]
                      = "cancel"]

Commit(r)  ==
          /\   rm_status[r] = "committing"
          /\   IF   isCommitFirstVal(r)        THEN
          /\   commitFirstVal(r)
          /\   rm_status '=[rm_status EXCEPT ![r]="committed"]
      ELSE
          /\   rm_status '=[rm_status EXCEPT![r]= "cancel"]
```

7. Finally, we define the Next operation in the process. Obviously, it must complete four processes in sequence, as shown in the following TLA+:

```
Next ==
      \E  r \in RM:
      Begin(r) \/ Loading(r) \/ preCommit(r) \/ commit(r)
```

We run the complete TLA+ program statement on TLC and get the result as shown in the Fig. 6.1. This algorithm generates a total of 1296 states, 324 different states have been found, and "no errors are found," which can verify the correctness of the algorithm SAOLA.

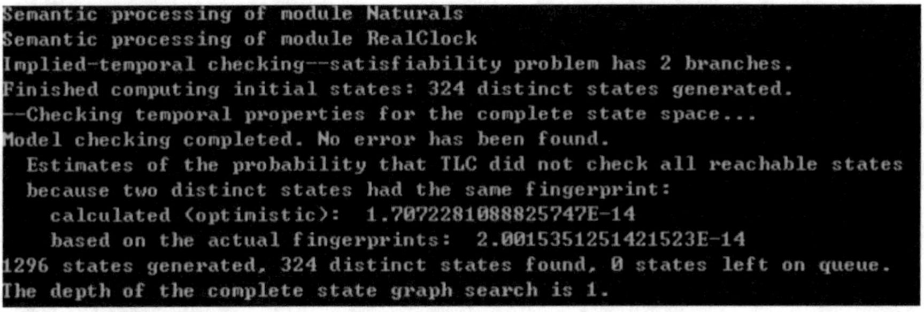

```
Semantic processing of module Naturals
Semantic processing of module RealClock
Implied-temporal checking--satisfiability problem has 2 branches.
Finished computing initial states: 324 distinct states generated.
--Checking temporal properties for the complete state space...
Model checking completed. No error has been found.
  Estimates of the probability that TLC did not check all reachable states
  because two distinct states had the same fingerprint:
    calculated (optimistic):  1.7072281088825747E-14
    based on the actual fingerprints:  2.0015351251421523E-14
1296 states generated, 324 distinct states found, 0 states left on queue.
The depth of the complete state graph search is 1.
```

Fig. 6.1 TLA+ result with running in TLC

6.5   Evaluation

6.5.1   Experimental Setup

Unless otherwise mentioned, all experiments are conducted on the Kodiak test bed. Each machine has a single core 2.7 GHz Intel Core i5 with 8GB RAM and 500GB SSD. Most experiments are bottlenecked on the server CPU. We have achieved much higher throughput when running on a local testbed with faster CPUs.

6.5.2   Experimental Case

We evaluate 2PC+'s performance under the data consistency of the Ctrip e-commerce platform (MSECP). As shown in the Fig. 6.2 below, the experimental case of MSECP contains a total of three micro-service modules: COMS, CSMS, and AMS. In the beginning, the customer initiates a request to create a new order. As the initiator of COMS, RPC remotely calls the gateways in CSMS and AMS to complete the order delivery and deduction. The CSMS and AMS then return the results of the operations in the respective service units. Finally, COMS returns the final response to the customer based on the returned results. If the operation is all successful, the order is created successfully. Otherwise, the order creation fails and the data is rolled back immediately. The MSECP is based on the spring cloud micro-services framework [24]. Finally, we deploy three micro-service module clusters.

6.5.3   RT Experiment

The SAOLA algorithm in 2PC+ needs to determine the performance improvement after optimization based on the response time (RT) of the service. In this test experiment, the createOrder interface was called concurrently with 10, 20, 50, 100,

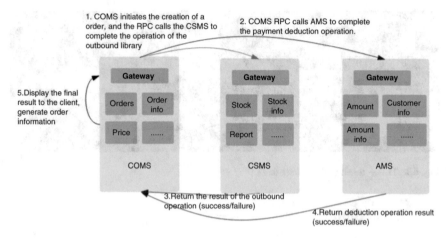

Fig. 6.2 A case of Ctrip MSECP

200, 300, and 500 sets of threads. A result of ten sets of comparison experiments was performed to calculate the RT mean values. The experimental results are shown in Fig. 6.2.

According to the analysis of the experimental results, when the number of threads concurrency is between 10 and 50, 2PC+ does not show a significant advantage on RT. However, when the thread concurrency reaches 100, it is shortened from the original 241.7ms to 83.4ms, which is 34.5% of the original RT duration. As thread concurrency continues to grow, the RT performance of the optimization scheme becomes more apparent. When the number of thread concurrency reaches 300, the RT value of the original 2PC is 820.5ms, and the RT value of 2PC+ is only 217.6ms, which is 26.5% of 2PC time. When the number of thread concurrent requests reaches 500, the 2PC's RT value is 1357.8ms, and 2PC+ is shortened to 473.6ms, which is 34.8% of the original time.

In summary, in the higher concurrent thread request scenario, the optimization algorithm in 2PC+ performs well in RT performance. Compared with 2PC, the RT performance is improved by 2.87 times to 3.77 times.

6.5.4 TPS Experiment

Similarly, transactions per second (TPS) is also one of the indicators for evaluating performance. The experimental results are shown in Fig. 6.3 below.

Through the experimental results, it can be known that when the number of concurrent threads is less than 50, 2PC+ has no obvious advantage in TPS performance. However, as the concurrency of threads increases, the TPS performance advantages of the optimization scheme gradually emerge. When the number of concurrent threads is between 100 and 200, 2PC+ can be maintained at TPS between 627.0 and

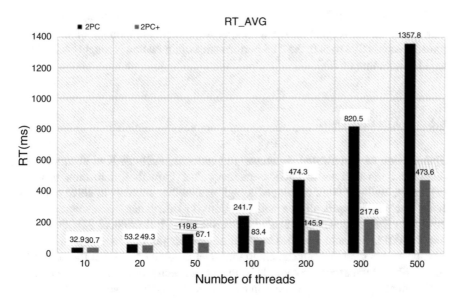

Fig. 6.3 RT experiment comparison result

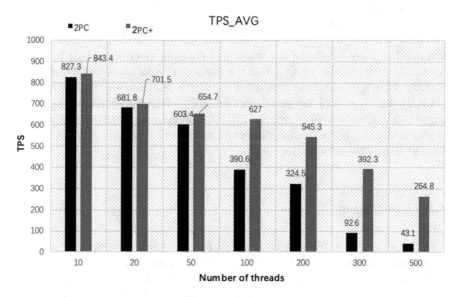

Fig. 6.4 TPS experiment comparison result

545.3, which is about 60.5% to 68.0% higher than that of 2PC's 390.6 and 324.5. When the number of concurrent threads reaches 300 to 500, the TPS of 2PC+ can be maintained at 392.3 to 264.8, which is 323.7% to 514.4% higher than the TPS of 92.6 and 43.1 in 2PC (Fig. 6.4).

As a summary, 2PC+ performs equally effective in TPS performance in higher concurrent thread request scenarios. Especially when the number of concurrent threads reaches 300 to 500, the TPS performance optimization is roughly 4.24 times to 6.14 times that of 2PC.

6.5.5   Related Work

Transaction of distributed database In academia, Kallman R et al. implement distributed transactions in H-Store memory data through serialization methods [9]. Later, Bailis P et al. developed a high availability solution to solve common faults such as network delays and partitions in distributed transactions [9]. In 2014, Mu S et al. developed a more concurrency control protocol ROCOCO [10] based on 2PL protocol and OCC, which has higher performance in dealing with distributed transaction conflicts. Guerraoui R et al. pointed out that the core rules for distributed transaction commit [11], that is, atomicity, must meet the conditions for the final agreement of all nodes in a distributed system. In the industry, Internet companies including Google, eBay, Alibaba, and PingCAP have been developing transaction solutions in distributed systems in recent years. In 2012, Google released the Spanner distributed database [12] and then in 2017 released the world's first commercial cloud data support for distributed transactions Cloud Spanner. Since then, Alibaba has developed the distributed database OceanBase [13] based on Spanner's design ideas, which solves the problem of data consistency and cross-database table transactions in distributed systems. It has extremely high processing performance. PingCAP [14] also released a distributed database VoltDB [15], which supports horizontal elastic extension, ACID transaction, standard SQL, and MySQL syntax and protocol, with high data consistency and high availability, and can support distributed transactions.

6.6   Conclusion

This paper presented 2PC+, a novel concurrency control protocol for distributed transactions in micro-service architecture. 2PC+ optimizes the synchronization blocking situation of transaction threads and reduces the probability of conflict between transactions due to high concurrency in micro-service architecture. And through the specific experimental data verification, compared to 2PC, 2PC+ has more efficient performance in RT and TPS.

Acknowledgment Our deepest gratitude goes to the anonymous reviewers for their valuable suggestions to improve this paper. This paper is partially supported by funding under National Key Research and Development Project 2017YFB1001800, NSFC Project 61972150, and Shanghai Knowledge Service Platform Project ZF1213.

References

1. B. Familiar, Microservice architecture [J] (2015)
2. F. Rademacher, S. Sachweh, A. Zündorf, Analysis of service-oriented modeling approaches for viewpoint- specific model-driven development of microservice architecture [J] (2018)
3. B.W. Lampson, D.B.A Lomet, New presumed commit optimization for two phase commit [C]. in *International Conference on Very Large Data Bases* (1993)
4. Qi Z, Xiao X, Zhang B, et al. Integrating X/Open DTP into Grid services for Grid transaction processing [C]. in *IEEE International Workshop on Future Trends of Distributed Computing Systems* (2004)
5. D. Peng, F. Dabek, Large-scale incremental processing using distributed transactions and notifications. in *Operating Systems Design and Implementation* (Oct. 2010)
6. A Dey, A Fekete, R Nambiar, et al., YCSB+T: Benchmarking web-scale transactional databases [C]. in *IEEE International Conference on Data Engineering Workshops* (2014)
7. S.J. Mullender, A.S. Tanenbaum, A distributed file service based on optimistic concurrency control [M]. *ACM SIGOPS Operating Systems Review* (2017), pp. 51–62
8. T. Neumann, T. Mühlbauer, A. Kemper, Fast serializable multi-version concurrency control for main-memory database systems [C].in *ACM Sigmod International Conference on Management of Data* (2015)
9. P. Bailis, A. Davidson, A. Fekete, et al., Highly available transactions: Virtues and limitations (extended version) [J]. Proc. Vldb Endowment **7**(3), 181–192 (2013)
10. S Mu, Y Cui, Y Zhang, et al., Extracting more concurrency from distributed transactions [C]. in *Usenix Conference on Operating Systems Design & Implementation* (USENIX Association, 2014)
11. R. Guerraoui, J Wang, How fast can a distributed transaction commit [C]. in *The 36th ACM SIGMOD-SIGACT-SIGAI Symposium* (ACM, 2017)
12. J.C. Corbett, J. Dean, M. Epstein, Spanner: Google's globally-distributed database [C]. in *Proceedings of OSDI* (2012)
13. OceanBase. https://oceanbase.alipay.com/
14. TiDB. https://pingcap.com/
15. D. Bernstein, Today's Tidbit: VoltDB [J]. IEEE Cloud Comput. **1**(1), 90–92 (2014)

Index

© Springer Nature Switzerland AG 2021
H. Gao, Y. Yin (eds.), *Intelligent Mobile Service Computing*, EAI/Springer
Innovations in Communication and Computing,
https://doi.org/10.1007/978-3-030-50184-6

Printed in the United States
by Baker & Taylor Publisher Services